SpringerBriefs in Molecular Science

SpringerBriefs in Molecular Science present concise summaries of cutting-edge research and practical applications across a wide spectrum of fields centered around chemistry. Featuring compact volumes of 50 to 125 pages, the series covers a range of content from professional to academic. Typical topics might include:

- A timely report of state-of-the-art analytical techniques
- A bridge between new research results, as published in journal articles, and a contextual literature review
- A snapshot of a hot or emerging topic
- An in-depth case study
- A presentation of core concepts that students must understand in order to make independent contributions

Briefs allow authors to present their ideas and readers to absorb them with minimal time investment. Briefs will be published as part of Springer's eBook collection, with millions of users worldwide. In addition, Briefs will be available for individual print and electronic purchase. Briefs are characterized by fast, global electronic dissemination, standard publishing contracts, easy-to-use manuscript preparation and formatting guidelines, and expedited production schedules. Both solicited and unsolicited manuscripts are considered for publication in this series.

Manish Kumar · Nilotpal Das · Kanika Dogra

Endocrine Disruptors

Environmental Interactions, Occurrences, and Toxicological Aspects

 Springer

Manish Kumar
Escuela de Ingeniería y Ciencias
School of Engineering and Science
Tecnológico de Monterrey
Monterrey, Nuevo León, Mexico

Nilotpal Das
ENCORE Insoltech Consulting
Gandhinagar, Gujarat, India

Kanika Dogra
Sustainability Cluster
School of Advanced Engineering
University of Petroleum and Energy Studies
Dehradun, Uttarakhand, India

ISSN 2191-5407   ISSN 2191-5415 (electronic)
SpringerBriefs in Molecular Science
ISBN 978-3-032-15156-8   ISBN 978-3-032-15157-5 (eBook)
https://doi.org/10.1007/978-3-032-15157-5

This Springer imprint is published by the registered company Springer Nature Switzerland AG
The registered company address is: Gewerbestrasse 11, 6330 Cham, Switzerland

If disposing of this product, please recycle the paper.

Dedicated to Prof. Ram Sharma—whose sense of duty, depth of dedication, and debonair grace have defined inspired leadership as Vice-Chancellor of UPES, Dehradun, India

Foreword

Endocrine-disrupting chemicals (EDCs) have emerged as one of the most critical classes of environmental contaminants, posing significant threats to global public health, biodiversity, and ecosystem integrity. These compounds can interfere with hormonal regulation, resulting in profound physiological, developmental, and behavioural effects in humans and animals. The diversity of their sources, their environmental persistence, and the complexity of their interactions within biological systems have made EDCs a formidable scientific and regulatory challenge of the twenty-first century.

This book titled *"Endocrine Disruptors: Environmental Interactions, Occurrences, and Toxicological Aspects"* presents a comprehensive and interdisciplinary examination of endocrine-disrupting chemicals, their origins, classification, environmental fate, exposure pathways, toxicological mechanisms, and mitigation strategies. While risks posed by individual compounds is documented earlier, mounting evidence now highlights the cumulative and synergistic impacts of multiple EDCs, often described as the "cocktail effect." Such interactions may amplify adverse biological responses even at trace concentrations, emphasizing the need for a mechanistic understanding of EDC behaviour across diverse biotic and abiotic environmental matrices.

This book will serve as an introductory reference as well as an advanced research resource for undergraduate and graduate students, researchers, and professionals working in environmental science, toxicology, and public health. By integrating contemporary scientific findings with regulatory and policy perspectives, this book will bridge the gap between environmental chemistry and human health, while promoting evidence-based strategies to minimize EDC exposure and enhance environmental resilience.

Addressing EDC pollution requires coordinated global action, combining scientific innovation, public awareness, effective policy implementation, and change in mind set of the process engineers to use environmentally benign compounds in the product formulation to protect both present and future generations from the persistent and long-lasting impacts of endocrine disruption. I complement the authors for giving fresh insights and holistic information on adverse effects of EDC. I urge the readers to go judgmentally through the chapters of this book to capture prevailing knowledge and develop your own policy and thinking for curbing pollution caused by EDCs to protect the environment.

Karaikal, India   Makarand M. Ghangrekar
January, 2026

Preface

Endocrine-disrupting chemicals (EDCs) have emerged as one of the most critical classes of environmental pollutants threatening global health, biodiversity, and ecosystem integrity. These compounds, both synthetic and naturally occurring, can interfere with hormonal regulation, leading to profound physiological, developmental, and behavioral changes in humans and wildlife. The complexity of their sources, persistence, and interactions has made EDCs a pressing scientific and regulatory challenge in the twenty-first century.

This book aims to present a comprehensive and interdisciplinary understanding of EDCs by examining their origins, types, environmental fate, exposure routes, toxicological mechanisms, and potential remediation strategies. While considerable research has highlighted the hazards of individual chemicals, growing evidence now points to the cumulative and synergistic effects of multiple EDCs, the so-called "cocktail effect," which can amplify biological risks even at trace concentrations. These interactions necessitate a deeper exploration of the mechanisms governing EDC behavior in diverse environmental matrices, including water, soil, air, and living systems.

The chapters in this volume provide an integrated overview of current global knowledge on endocrine disruptors. Chapter 1 introduces the concept of EDCs, tracing their historical background, classification, and implications for human health and environmental sustainability, emphasizing their connection to the United Nations Sustainable Development Goals (SDGs 3 and 6). Chapter 2 discusses the worldwide occurrence of EDCs across various environmental compartments and summarizes the evolving international guidelines and regulatory frameworks in different regions, including the United States, the European Union, Asia, Africa, and Oceania. Chapter 3 delves into the pathways of exposure, metabolic transformations, and toxicological effects, presenting a mechanistic understanding of how EDCs interact within biological systems. Finally, Chapter 4 explores the interactions of EDCs with organisms and environmental processes, as well as the advances in physical, chemical, and biological methods for their degradation and removal.

This book has been designed to serve as both a foundational reference and an advanced research guide for students, academics, and practitioners in environmental science, toxicology, public health, and sustainability studies. By bringing together insights from recent scientific literature and regulatory perspectives, it aspires to bridge the gap between environmental chemistry and human health, encouraging evidence-based strategies to minimize EDC exposure and promote global environmental resilience.

Ultimately, the goal of this book is not only to document the growing body of knowledge on endocrine-disrupting chemicals but also to inspire a forward-looking vision for sustainable management practices. Addressing EDCs pollution requires coordinated global action spanning scientific innovation, public awareness, and policy reform to safeguard both present and future generations from their pervasive and lasting effects.

Monterrey, Mexico   Manish Kumar
Gandhinagar, India   Nilotpal Das
Dehradun, India   Kanika Dogra

Competing Interests The authors have no competing interests to declare that are relevant to the content of this manuscript.

Contents

Chapter 1
Understanding Endocrine Disrupting Chemicals: History, Classification, and Global Perspectives

Abstract Endocrine-disrupting chemicals (EDCs) encompass a wide range of natural and man-made substances that disrupt hormonal functions in humans and animals, resulting in reproductive, metabolic, cognitive, and cancer-related consequences. Since the mid-twentieth century, their extensive use in plastics, pesticides, firefighting foams, and consumer goods has led to a lasting global environmental presence and bioaccumulation. Historically, awareness developed from mass production in the 1950s to global regulation via key events like Silent Spring, the Stockholm Convention, and the REACH framework. Key EDC categories encompass pesticides, bisphenols, phthalates, PFAS, and trace elements, several of which are designated as persistent organic pollutants (POPs). These substances enter the ecosystem through industrial waste, agricultural runoff, and household waste, exposing people mainly through food, air inhalation, and skin contact. Growing evidence suggests that the combined or "cocktail" effects of various EDCs enhance toxicological results beyond those seen with exposure to individual compounds. Although there is increasing global awareness, regulatory inconsistencies remain, especially in developing areas that have restricted oversight capabilities. Filling these gaps is essential for promoting the United Nations Sustainable Development Goals (SDG 3: Good Health and Well-being; SDG 6: Clean Water and Sanitation). This research combines historical, toxicological, and policy viewpoints to emphasize the immediate necessity for unified, science-driven regulation of EDCs globally.

Keywords PFAS · Phthalates · Pesticides · Environmental exposure · Hormonal disruption · Sustainable development goals · Human health

1.1 Introduction

The term endocrine-disrupting chemicals (EDCs) refers to a class of both naturally occurring and man-made compounds that have the potential to influence or interfere with hormonal actions in both human and non-human organisms, resulting in negative health outcomes (La Merrill et al., 2020), cognitive defects (Govarts et al., 2012; Kahn

M. Kumar et al., *Endocrine Disruptors*,
SpringerBriefs in Molecular Science, https://doi.org/10.1007/978-3-032-15157-5_1

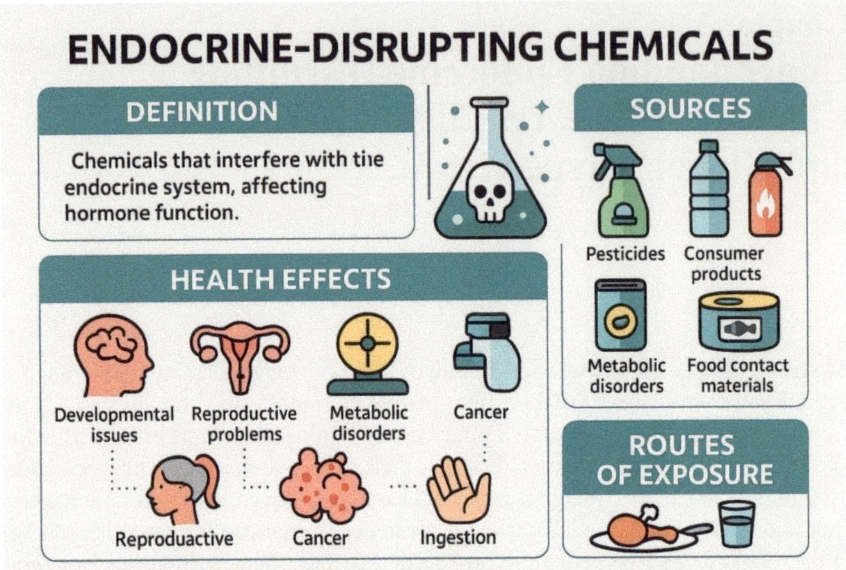

Fig. 1.1 Key aspects of EDCs include their definition, major sources, associated health effects, and primary routes of human exposure. The visual emphasizes the widespread presence and potential risks of EDCs in the environment and public health

et al., 2020; Pinson et al., 2016; Zuo et al., 2024), reproductive problems (Balabanič et al., 2011; Hall et al., 2025; Hassan et al., 2024), metabolic diseases and disorders (Audouze et al., 2020; Encarnação et al., 2019), and a variety of tumors and cancers (Macedo et al., 2023; Rodgers et al., 2018) as shown in Fig. 1.1. The European Commission states that three essential steps can determine if a chemical qualifies as an EDC: (EFSA, 2013; Lauretta et al., 2019; Macedo et al., 2023; Slama et al., 2016) (1) endocrine activity, (2) harmful and/or pathologic endocrine-mediated activity, and (3) the cause-and-effect link between substance and endocrine activity in exposed people.

1.1.1 Historical Background and Emergence of EDCs

EDCs became popular in the latter half of the twentieth century, especially post-World War II, due to their numerous applications in various aspects of life, i.e., plasticizers, pesticides, surfactants, and firefighting foams. The chronology tracks their development historically from synthesis and extensive application during the early 1800s–1950s, such as PVC, DDT, PCBs, and BPA, to increasing awareness of their dangers and early regulation between 1950 and 1980, which was punctuated by milestones like Silent Spring, the Kepone spill, and the prohibition of PCBs. During 1980–1999, evidence grew from science in the form of coining the term "endocrine," the discovery of leaching of BPA, and the first international conference on EDCs. During 1999–2007, policy action at a global level increased with the Stockholm Convention, the Hormone Disruption Research Act, coining the term "microplastics," and correlations between EDCs, obesity, and diabetes. The 2007–2015 era saw intensified regulation, including the implementation of REACH, Canada's declaration of BPA as a hazardous substance, EU prohibition of BPA in baby bottles, and the WHO/ACOG reports regarding EDCs. The most recent era (2015–2025) sees increased global awareness by means of activities like FIGO and the U.S. DOHaD Society, growing focus post-Dark Waters on PFAS, the EU-wide prohibition of BPA, and continued demands for swift regulatory changes to treat EDC exposure and persistence (HEEDS) (Fig. 1.2). According to Eurostat (2020) and Macedo et al. (2023), the European Union (EU) produced around 211 million tons of these chemicals in 2019. Over 1000 chemicals released or existing in the environment have been classified as EDCs, and the number is constantly rising (Macedo et al., 2023). According to the Fourth National Report on Human Exposure to Environmental Chemicals, practically all Americans, regardless of age (6 years or older), have detectable amounts of common EDCs in their urine, including phthalate metabolites, bisphenol A, and triclosan. Current scientific bodies have suggested lowering exposure to pollutants that change hormones, including the United Nations Environment Program (UNEP), the World Health Organisation (WHO), the Endocrine Society, and the American Academy of Paediatrics (ESAAP) (Wong et al., 2017).

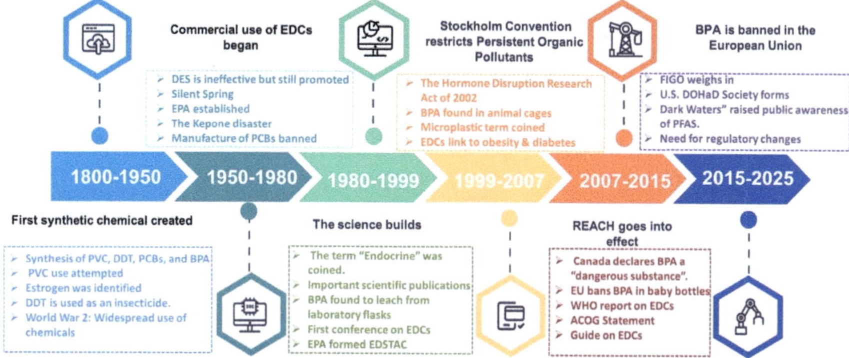

Fig. 1.2 Timeline illustrating the evolution of endocrine-disrupting chemicals (EDCs) from their synthesis and widespread use to scientific recognition, regulatory milestones, and global awareness. Key events include early use of PVC, DDT, PCBs, and BPA; publication of Silent Spring; the Stockholm Convention; REACH implementation; and recent actions such as the EU ban on BPA and heightened awareness of PFAS

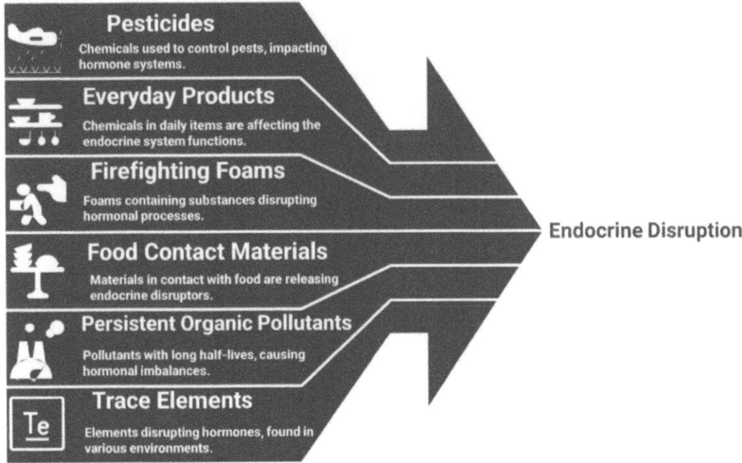

Fig. 1.3 Major categories of endocrine-disrupting chemicals (EDCs) and their sources. These include pesticides, everyday consumer products, firefighting foams, food contact materials, persistent organic pollutants (POPs), and trace elements, all of which can interfere with hormonal processes and contribute to endocrine disruption

1.2 Classification and Major Types of EDCs

According to Macedo et al. (2023), EDCs are categorized as follows: (a) pesticides (chemicals like DDT that kill and control pests); (b) chemicals associated with or found in everyday products like electronics, water bottles, textiles and clothing, pharmaceuticals, and personal care products (PPCPs); firefighting foams (e.g., phthalates, Bisphenols, per- and polyfluoroalkyl substances (PFASs)); and food contact materials (e.g., bisphenols). The Stockholm Convention has also identified some of them as persistent organic pollutants (POPs), including both new additions like perfluorooctanoic acid (PFOA) and perfluorooctane sulfonic acid (PFOS), and DDT (UNEP, 2022). In fact, because of their lengthy half-lives and environmental persistence, PFASs (which include PFOS and PFOA) are also part of EDCs substances that have been dubbed "forever chemicals." Trace elements such as aluminum (Al), arsenic (As), cadmium (Cd), copper (Cu), lead (Pb), mercury (Hg), manganese (Mn), zinc (Zn), and others have also been shown in several studies to have the ability to disrupt hormones (Kilic et al., 2018; Sabir et al., 2019) (Fig. 1.3).

1.3 Health Effects and Toxicological Implications

EDCs, being hydrophobic, are present in water together with particulate matter and thus settle in the bottom sediments, where they accumulate continuously through electrostatic interactions (Rogers et al., 2013). Besides WWTP discharges, other sources include animal feeding operations, land-applied biosolids, row-crop production, on-site wastewater disposal systems, recreational activities, transportation, and wash-off from roadways, as well as atmospheric deposition. Apart from this, EDCs are introduced in aquatic environments through households, agriculture, and industrial release of leftover chemicals (Qin et al., 2012; Zimmer et al., 2010). Therefore, humans and other species are always in danger of exposure. EDCs can be exposed through three main routes: breathing, direct contact, or cutaneous (dermal/skin), and diet (food, water, pesticides, food packaging materials, etc.). Of these three, dietary exposure has been deemed the most significant (Lauretta et al., 2019; Macedo et al., 2023). Different researchers, such as Cruz et al. (2025) and Amine et al. (2025), have speculated that exposure to these chemicals may contribute to the rising prevalence of conditions such as diabetes, obesity, cognitive impairment, neurodegenerative diseases, early puberty, thyroid disorders, heart diseases, and infertility. One of the significant concerns regarding EDCs is their persistence and bioaccumulation in biological tissues. Many persistent organic pollutants (POPs) exhibit lipophilic properties, enabling their accumulation in both human and non-human animal tissues (Gore et al., 2015; Heindel et al., 2015).

1.4 Global Relevance and Link to Sustainable Development Goals

Recent research has underscored the growing importance of global regulatory interventions in mitigating exposure to EDCs. The European Union, through the Registration, Evaluation, Authorisation and Restriction of Chemicals (REACH) regulation, has implemented stringent measures aimed at identifying and phasing out substances with endocrine-disrupting properties. Similarly, countries such as Canada and Japan are revising their chemical risk assessment frameworks to include endocrine-related mechanisms and chronic exposure scenarios.

Despite these advances, significant regulatory disparities persist across regions. Many developing nations, particularly in Latin America and parts of Asia, still lack comprehensive frameworks or standardized guidelines to assess and control EDCs exposure. This global inconsistency highlights the urgent need for international collaboration and harmonization of toxicological and regulatory standards to ensure equitable protection, especially for vulnerable populations such as children and pregnant women, who are biologically more susceptible to endocrine disruption (European Commission, 2023).

Another critical concern is the combined or "cocktail effect" arising from prolonged exposure to multiple EDCs through various environmental matrices over a lifetime. These mixtures can interact synergistically or cumulatively within biological systems, intensifying adverse outcomes beyond the effects of individual compounds (Buoso et al., 2020; Macedo et al., 2023). Such interactions complicate risk assessment processes and underscore the necessity of adopting mixture-based evaluation models in toxicological research and policymaking. In the broader sustainability context, the persistence and bioaccumulative nature of EDCs directly threaten global efforts to achieve several United Nations Sustainable Development Goals (SDGs), notably SDG 6 (Clean Water and Sanitation) and SDG 3 (Good Health and Well-being). The continued release and environmental accumulation of EDCs undermine progress toward these objectives by compromising water quality, ecosystem health, and human well-being.

Therefore, this study seeks to critically analyze both historical and contemporary findings to provide a comprehensive understanding of EDCs encompassing their occurrence across environmental matrices, global distribution patterns, molecular interactions, and toxicological implications. Furthermore, it emphasizes the need for intensified research into the combined effects of multiple EDCs and advocates for the establishment of robust, science-based regulatory frameworks to mitigate their impact on public health, ecosystems, and global sustainability goals.

References

Amine, I., Guillien, A., Bayat, S., Lyon-Caen, S., Ouidir, M., Sabaredzovic, A., Sakhi, A.K., Thomsen, C., Valmary-Degano, S., Philippat, C., & Siroux, V. (2025). Early-life exposure to mixtures of endocrine-disrupting chemicals and a multi-domain health score in preschool children. *Environmental Research*, 121173.

Audouze, K., Sarigiannis, D., Alonso-Magdalena, P., Brochot, C., Casas, M., Vrijheid, M., Babin, P. J., Karakitsios, S., Coumoul, X., & Barouki, R. (2020). Integrative strategy of testing systems for identification of endocrine disruptors inducing metabolic disorders—An introduction to the OBERON project. *International Journal of Molecular Sciences, 21*(8), 2988.

Balabanič, D., Rupnik, M., & Klemenčič, A. K. (2011). Negative impact of endocrine-disrupting compounds on human reproductive health. *Reproduction, Fertility and Development, 23*(3), 403–416.

Buoso, E., Masi, M., Racchi, M., & Corsini, E. (2020). Endocrine-disrupting chemicals'(EDCs) effects on tumour microenvironment and cancer progression: Emerging contribution of RACK1. *International Journal of Molecular Sciences, 21*(23), 9229.

Cruz, J. C., Rocha, B. A., Souza, M. C. O., Kannan, K., & Júnior, F. B. (2025). Co-exposure to multiple endocrine-disrupting chemicals and oxidative stress: Epidemiological evidence of nonmonotonic dose response curves. *Science of the Total Environment, 969*, 178952.

EFSA Scientific Committee (2013). Scientific Opinion on the hazard assessment of endocrine disruptors: Scientific criteria for identification of endocrine disruptors and appropriateness of existing test methods for assessing effects mediated by these substances on human health and the environment. *EFSA Journal, 11*(3), 3132.

Encarnação, T., Pais, A. A., Campos, M. G., & Burrows, H. D. (2019). Endocrine disrupting chemicals: Impact on human health, wildlife and the environment. *Science Progress, 102*(1), 3–42.

European Commission: Sustainable chemicals: new rules to identify endocrine disruptors and long-lasting chemicals enter into force. (2023). Accessed May 10, 2025.

Eurostat (2020). Chemicals production and consumption statistics [WWW Document]. https://ec.europa.eu/eurostat/statisticsexplained/index.php?title=Chemicals_production_and_consumption_statistics#Production_of_chemicals_hazardous_to_health.

Gore, A. C., Chappell, V. A., Fenton, S. E., Flaws, J. A., Nadal, A., Prins, G. S., Toppari, J. & Zoeller, R. T. (2015). EDC-2: the endocrine society's second scientific statement on endocrine-disrupting chemicals. *Endocrine Reviews*, *36*(6), E1-E150.

Govarts, E., Nieuwenhuijsen, M., Schoeters, G., Ballester, F., Bloemen, K., De Boer, M., Chevrier, C., Eggesbø, M., Guxens, M., Krämer, U., & Legler, J. (2012). Birth weight and prenatal exposure to polychlorinated biphenyls (PCBs) and dichlorodiphenyldichloroethylene (DDE): A meta-analysis within 12 European Birth Cohorts. *Environmental Health Perspectives*, *120*(2), 162–170.

Hall, J. M. (2025). Endocrine disruptors in molecular and structural endocrinology. *Frontiers in Endocrinology*, *16*, 1561253.

Hassan, S., Thacharodi, A., Priya, A., Meenatchi, R., Hegde, T. A., Thangamani, R., Nguyen, H. T., & Pugazhendhi, A. (2024). Endocrine disruptors: Unravelling the link between chemical exposure and women's reproductive health. *Environmental Research*, *241*, 117385.

Heindel, J. J., Vom Saal, F. S., Blumberg, B., Bovolin, P., Calamandrei, G., Ceresini, G., Cohn, B. A., Fabbri, E., Gioiosa, L., Kassotis, C., & Legler, J. (2015). Parma consensus statement on metabolic disruptors. *Environmental Health*, *14*, 1–7.

Kahn, L. G., Philippat, C., Nakayama, S. F., Slama, R., & Trasande, L. (2020). Endocrine-disrupting chemicals: Implications for human health. *The Lancet Diabetes & Endocrinology*, *8*(8), 703–718.

Kilic, S., Tongur, T., Kilic, M., & Erkaymaz, T. (2018). Determination of some endocrine-disrupting metals and organochlorinated pesticide residues in baby food and infant formula in Turkish markets. *Food Analytical Methods*, *11*, 3352–3361.

La Merrill, M. A., Vandenberg, L. N., Smith, M. T., Goodson, W., Browne, P., Patisaul, H. B., Guyton, K. Z., Kortenkamp, A., Cogliano, V. J., Woodruff, T. J., & Rieswijk, L. (2020). Consensus on the key characteristics of endocrine-disrupting chemicals as a basis for hazard identification. *Nature Reviews Endocrinology*, *16*(1), 45–57.

Lauretta, R., Sansone, A., Sansone, M., Romanelli, F., & Appetecchia, M. (2019). Endocrine disrupting chemicals: Effects on endocrine glands. *Frontiers in Endocrinology*, *10*, 178.

Macedo, S., Teixeira, E., Gaspar, T. B., Boaventura, P., Soares, M. A., Miranda-Alves, L., & Soares, P. (2023). Endocrine-disrupting chemicals and endocrine neoplasia: A forty-year systematic review. *Environmental Research*, *218*, 114869.

Pinson, A., Bourguignon, J. P., & Parent, A. S. (2016). Exposure to endocrine disrupting chemicals and neurodevelopmental alterations. *Andrology*, *4*(4), 706–722.

Qin, M., Yang, H., Chen, S., Xie, H., & Guan, J. (2012). Photochemical characteristics of diclofenac and its photodegradation of inclusion complexes with β-cyclodextrins. *Quimica Nova*, *35*, 559–562.

Rodgers, K. M., Udesky, J. O., Rudel, R. A., & Brody, J. G. (2018). Environmental chemicals and breast cancer: An updated review of epidemiological literature informed by biological mechanisms. *Environmental Research*, *160*, 152–182.

Rogers, J. A., Metz, L., & Yong, V. W. (2013). Endocrine disrupting chemicals and immune responses: A focus on bisphenol-A and its potential mechanisms. *Molecular Immunology*, *53*(4), 421–430.

Sabir, S., Akhtar, M. F., & Saleem, A. (2019). Endocrine disruption as an adverse effect of non-endocrine targeting pharmaceuticals. *Environmental Science and Pollution Research*, *26*, 1277–1286.

Slama, R., Bourguignon, J. P., Demeneix, B., Ivell, R., Panzica, G., Kortenkamp, A., & Zoeller, R. T. (2016). Scientific issues relevant to setting regulatory criteria to identify endocrine-disrupting substances in the European Union. *Environmental Health Perspectives*, *124*(10), 1497–1503.

UNEP (2022). The new POPs under the Stockholm convention. Secretariat of the Stockholm Convention. Retrieved January 8, 2025, from http://chm.pops.int/TheConvention/ThePOPs/The NewPOPs/tabid/2511/Default.aspx.

Wong, K. H., & Durrani, T. S. (2017). Exposures to endocrine disrupting chemicals in consumer products—A guide for pediatricians. *Current Problems in Pediatric and Adolescent Health Care, 47*(5), 107–118.

Zimmer, M. (2010). Glowing genes: A revolution in biotechnology. Prometheus Books.

Zuo, Q., Gao, X., Fu, X., Song, L., Cen, M., Qin, S., & Wu, J. (2024). Association between mixed exposure to endocrine-disrupting chemicals and cognitive function in elderly Americans. *Public Health, 228,* 36–42.

Chapter 2
Global Occurrence and Guidelines of Endocrine-Disrupting Chemicals

Abstract Environmental pollutants known as EDCs are present in drinking water, food, consumer goods, and the atmosphere and pose serious threats to ecosystems and human health. These pollutants come from industrial processes, wastewater discharge, and the usage of consumer products. They include hormones, per- and polyfluoroalkyl substances (PFAS), bisphenol A (BPA), phthalates, polycyclic aromatic hydrocarbons (PAHs), and pesticides. This study examines the global occurrence, sources, and fate of EDCs and evaluates the regulatory frameworks. Excess hormone levels are found in waters in Africa and Europe, and Africa suffers from higher contamination levels due to inadequate wastewater treatment. PFAS exceed regulatory standards in Europe, the Americas, and Asia, while BPA and phthalates persist in bottled water, rivers, and food, raising concerns over human exposure. The USA uses TSCA and EDSP, but loopholes such as the GRAS exemption under the FD&C Act permit unregulated EDC use. The EU uses a precautionary strategy, incorporating EDC legislation into REACH, the Plant Protection Products Regulation, and BPR. Asia, Africa, and Oceania exhibit mixed regulatory strategies, with some countries having strict bans, while others use risk-based strategies.

Keywords Occurrence · Guidelines · Surface water · Atmosphere · Groundwater

2.1 Introduction

Endocrine-disrupting chemicals represent a significant class of environmental pollutants due to their capacity to interfere with hormonal regulation and cause adverse effects in various human organs, including breast tissue, the reproductive system, the pancreas, and adipose tissue. Both natural and synthetic EDCs such as endogenous hormones, pharmaceutical compounds, pesticides, alkylphenols (APs), bisphenol A (BPA), polyhalogenated compounds, and phthalates, have been introduced into ecosystems through diverse pathways. These pathways include contamination of surface water, groundwater, drinking water, wastewater, and sedimentary deposits. Natural sources, such as phytoestrogens, also contribute to the environmental burden

M. Kumar et al., *Endocrine Disruptors*,
SpringerBriefs in Molecular Science, https://doi.org/10.1007/978-3-032-15157-5_2

of EDCs (Metcalfe et al., 2022). Numerous studies have consistently detected EDCs at trace concentrations (ranging from ng/L to µg/L) across different global regions. Investigations in the United States (Zhang et al., 2007), Europe (Lindqvist et al., 2005), and Japan (Furuichi et al., 2004; Madsen et al., 2004) have elucidated the risks these substances pose to aquatic organisms and broader ecological systems. EDCs may exert a wide spectrum of toxicological effects, including carcinogenicity, genotoxicity, cytotoxicity, and neurotoxicity, all of which can negatively impact individual and population-level health and broader socioeconomic structures.

Despite growing scientific evidence revealing the extent and complexity of EDC contamination, the level of regulatory action and public engagement remains disproportionately low. A heightened global commitment is essential to address the multifaceted challenges posed by EDCs. This chapter elucidates the worldwide prevalence of EDCs, with particular stress on their general abundance in numerous environmental matrices like surface water, groundwater, soil, and biota. The chapter also provides an overview of contemporary national and international regulations concerning EDCs, sketching regulatory frameworks, tolerable limits, and monitoring difficulties in detecting and controlling them.

2.2 Global Occurrence of EDCs

EDCs have been found in various matrices in the environment, from soils, sediments (Saha et al., 2022), seas, and brackish waters (Lestido-Cardama et al., 2023), surface and groundwater (Ghosh et al., 2024; Pignotti et al., 2017), as well as in different plants (Lee et al., 2022) and food materials (Mezcua et al., 2012; Ong et al., 2022). Table 2.1 summarizes the findings globally on occurrences of EDCs.

2.2.1 In Drinking Water, Surface, and Groundwater

The presence of hormones in surface freshwater was detected in various African and European nations at varying concentrations. The highest levels of the hormones were found in Africa, where there is the common discharge of untreated wastewater from homes and farm animals (Pironti et al., 2021). The concentrations range between 3000 and 20,000 times those found in Europe, with ranges of 3310–15,700 ng/L for 17β-estradiol and 510–45,500 ng/L for estriol (Olatunji et al., 2017). Estrone (E1) levels between 0.1 and 69 ng/L were recorded in France, the Czech Republic, Italy, Germany, Luxembourg, and Spain, while 0.23–13.7 ng/L of progesterone were recorded in France and Hungary (Fekadu et al., 2019). The EU has added E2 to the drinking water watch list and proposed maximum detection levels of 0.035 ng/L for EE2 and 0.4 ng/L for E1 and E2 to recommend monitoring for oestrogens present in surface water (Glineur et al., 2020).

Table 2.1 The global occurrence of different EDCs and their classes in various environmental matrices, including surface water (SW), groundwater (GW), wastewater (WWTPs), drinking water (DW), sediments, food, consumer products, and atmosphere. The concentration and detection frequency of EDCs are also mentioned. (*Note* All citations in this section have not been included to keep it concise)

Country & location	Chemicals	Occurrence	Concentration/detection frequency/observation trend	References
Australia (South East Queensland)	Phenols, hormones, and other EDCs	WWTPs	• Highest levels in influent: DEP: 1080, BP: 201; BBP: 134; DEHP: 716; OP: 229; NP: 3070; CP: 10.2; BPA: 140; Andr. Etio: 2040; E1: 268; α-E2: 13.1; β-E2: 16.6; E3: 110 ng/L, respectively • Highest levels in effluents: DEP: 4.9; BP: 34.3; BBP: 75.7; DEHP: 589; OP: 23.5; NP: 335; CP: 1.9; BPA: 86.7; Andr. Etio: BDL; E1: BDL; α-E2: 41.9, β-E2: 1.7; E3: 1.6 and BDL ng/L, respectively	Tan et al. (2007)
Austria	Phenols, and BPA	WWTPs	• Influents: BPA: 2376 ng/L; NP: 4031 ng/L, NP1EO: 7299 ng/L; NP2EO: 866 ng/L NP1EC: 737 ng/L; NP2EC: 840 ng/L; OP:680 ng/L; OP1EO: 660 ng/L; and OP2EO: 55 ng/L	Clara et al. (2005)
Brazil (Piracicaba River, São Paulo)	Hormones	SW, DW	• E1: 82 ng/L E2: 137 ng/L EE2: 480 ng/L E3: 26 ng/L	Torres et al. (2015)
Bangladesh (Mymensingh)	Arsenic	GW	• The highest level detected in a well at 251 μg/L	Ahmed et al. (2010)
Canada (Ontario)	Pharmaceuticals, hormones, and BPA	DWs	• BPA: Up to 87 ng/L; Naproxen: 199 ng/L; IBU: 79 ng/L; LIN: 143 ng/L; SMX: 284 ng/L; Monensin Na: 810 ng/L; APAP: 298 ng/L; TET: 35 ng/L	Kleywegt et al. (2011)

(continued)

Table 2.1 (continued)

Country & location	Chemicals	Occurrence	Concentration/detection frequency/observation trend	References
The Czech Republic	PFASs: 15 perfluoroalkyl acids	Consumer products	• Sample type: i) household equipment, ii) building materials, iii) car interior materials, and iv) WEEE. $\sum_{15}$PFAA (μg/kg) • Textiles: 77.64 μg/kg and Floor covering: 38.4 μg/kg • OSB wood: 18.3 μg/kg, and insulation materials: 34.3 μg/kg. Category 3 (35.5 μg/kg)	Bečanová et al. (2016)
France (Nice, Paris, Lille, Strasbourg, Lyon, Marseille, Toulouse, Bordeaux, Rennes)	Alkylphenol and Bisphenols	Bottle water, Raw water	• Bottled water—BPA: Up to 4210 ng/L BPA and 4-terBP: 247 ng/L • Raw water—BPA: Up to 1275 ng/L, BPF: 10 ng/L, NP: 170 ng/L, OP: < 17 ng/L, 4-terBP: 340 ng/L; NP1EC: 105 ng/L; NP1EO: 60 ng/L, NP2EO: 40 ng/L; OP2EO:20 ng/L • BPA: Up to 1430 ng/L; BPF:20 ng/L; NP: 605 ng/L; OP:130 ng/L; 4-terBP < 35 ng/L; NP1EC:615 ng/L; NP1EO: < 35 ng/L, NP2EO: 15 ng/L; and OP2EO: 10 ng/L	Colin et al. (2014)
	PPCPs, hormones, alkylphenols, phthalates and PFASs	Spring water and mineral water	• Out of 14,000 total bottled water samples, 99.7% below the LOQs, and • 87% of the LOQs were below10 ng/L	Le Coadou et al. (2017)
Germany (Berlin)	Phthalates and musk compounds	Atmosphere	• Indoor air—DBP: 1083 ng/m^3 (Highest mean level) • Musk compounds—HHCB: 101 ng/m3, AHTN: 44 ng/m^3 (Median values) • Households—AHTN: 83% and HHCB: 63% of the samples	Fromme et al. (2004)

(continued)

Table 2.1 (continued)

Country & location	Chemicals	Occurrence	Concentration/detection frequency/observation trend	References
Greece (Nafplio and Herakleio)	4-n-NP, NP1EO, TCS, BPA, and NP2EO	WWTPs	• Influents—4-n-NP: 1.04; NP1EO: 20.8; TCS: 23.9; BPA: 2.14; and NP2EO:13.4 μg/L, respectively • Effluents: 4-n-NP:0.9; NP1EO: 6.89; TCS: 6.88; BPA: 1.1; and NP2EO:17.4 μg/L, respectively	Stasinakis et al. (2008)
Global (India, China, Japan, New Zealand, the USA, Germany, Pacific Ocean, Atlantic Ocean, Indian Ocean, South China Sea, East China, North Japan Sea, Arctic and Antarctic regions)	Phytoestrogens	Food	• Isoflavones (β-glucosides, malonyl-β-glucoside, acetyl-β-glucoside and aglucons) found in high levels in members of the family Papilionoidea (legumes): soybean (*Glycine max*) and its products; chickpea (*Cicer arietinum*), alfalfa (*Medicago sativa*) and clover (Gen *Trifolium*), and some others like hops (*Humulus lupulus*)	Murphy and Hendrich (2002)
	Phytoestrogens (daidzein, genistein, secoisolariciresinol, and matairesinol)	Food	• Secoisolariciresinol was found to be very high in flaxseeds, sunflower seeds (*Helianthus*), caraway seeds (*Cumin cymicum*), cashew nuts (*Anacardium occidentale*), groundnuts (*Arachis hypogaea*), bramble (*Rubus fructicosus*), strawberry, lingonberry, and cranberry	Mazur et al. (2000)
	Phytoestrogens (coumestans)	Food	• Coumestans are present in plants of the families Leguminosae, Asteraceae, Mniaceae, Solanaceae, Mimosaceae, and Apocynaceae	Tu et al. (2021)
	AGE (advanced glycation end products)	Food	• Produced in various fast foods during extreme temperatures • Processing due to many processes like oxidation, glycoxidation, non-oxidative, hydration and dehydration, autoxidation, and fucosylation	Ravichandran et al. (2019)

(continued)

Table 2.1 (continued)

Country & location	Chemicals	Occurrence	Concentration/detection frequency/observation trend	References
	EDCs of various classes	Consumer products	• Parabens, formaldehyde, glutaraldehyde, aromatic amine derivatives, metal salts, bisphenols, TCS	Ripamonti et al. (2018)
	BPA	Atmosphere	• The highest levels of BPA were detected in India > USA > China > Japan > New Zealand • Chennai, India (17,400 pg/m^3 in PM$_{10}$), • Oceanic, Arctic, and Antarctic regions had the lowest levels	Fu and Kawamura (2010)
India (Mumbai, Delhi, Dehradun)	PAHs and pesticides	SW	• Overall levels, ΣPAHs was 157.96 ± 18.99 μgL^{-1}; • For carcinogenic PAHs (ΣC-PAHs): BaP: 81.31 ± 9.75 μgL^{-1}, up to 8.61 μgL^{-1} BkF: 24.7 μgL^{-1}; BbF: 15.91 μgL^{-1}; BghiP:15.40 μgL^{-1}; InP: 14.09 μgL^{-1} • Average levels of α- and β endosulfan: 137.75 ngL$^-$1; aldrin: 75.31 ngL^{-1}; dieldrin: 71.19 ngL$^-$1; endrin: 76.60 ngL^{-1}and chlorpyrifos: 208.77 ngL^{-1} (Vasai Creek)	Singare (2016)
	Arsenic	GW and Sediments	• The highest level was 22.1 μg L^{-1}	Das et al. (2017, 2018)
	OCPs, PCBs, PDBEs, dioxins, furans, PAEs and BPA	Food	• Higher food from the urban city of Delhi compared to the suburban city of Dehradun • Cottage cheese—PAEs: 665 ng/g and Potatoes—BPA: 73 ng/g; Fish—PAEs—477 ng/g, and BPA—16 ng/g	Sharma et al. (2021) and Chakraborty et al. (2022)

(continued)

Table 2.1 (continued)

Country & location	Chemicals	Occurrence	Concentration/detection frequency/observation trend	References
Italy Italian markets, River Tiber Basin	Phthalates and heavy metals	Food	• Coffee—DEHP: 0.16–1.87 μgmL^{-1} and DiBP: 0.01–0.36 μgmL^{-1} • Pb and Ni levels were found to exceed up to 79% (0.32–211.57 μg/dose) and 100% (166.25–1950.26 μg/dose) exceedance of the daily tolerable intake limits	De Toni et al. (2017)
	Natural and synthetic estrogens	WWTPs	• Inlet—E3: 80; E2: 12; E1: 52 and EE2: 3ngL^{-1} (Average)	Baronti et al. (2000)
Japan (Tohoku, Kansai, Tokyo)	Phenols, bisphenols, hormones, and other EDCs	WWTPs	• 14 chemicals were detected at levels above the minimum LODs in influents: 4-terBP, 4-n-octyl phenol, 4-tOP, NP, BPA, (2,4-dichlorophenol), DEP, DBP, DEHP, BBP, (benzo (a) pyrene di–Ethylhexyl adipate), benzophenone, (2,4,6-TPH of styrene trimers), n-butyl benzene	Nasu et al. (2001)
	Organophosphate and polybrominated flame retardants	Atmosphere	• OPCs:1260 ng/m^3 for TCPP (Max) • PBCs: 29.5 ng/m^3 for HBCD (Max)	Saito et al. (2007)
Norway (Country wide pilot study and Oslo)	PFASs	Consumer products	• Paints, AFFF, WPA, printed circuit boards, Coated Fabrics, and NSW had varying levels of PFASs; with AFFF documented to have high levels of FTSs, PFHxS, PFHpS, PFOS, PFDcS, PFBA, PFPA, and PFHxA	Herzke et al. (2012)
	Phthalates	Atmosphere	• Dominant phthalate was DBP in both PM$_{10}$ and PM$_{2.5}$ particles, with the major source possibly from rubber from tires	Rakkestad et al. (2007)

(continued)

Table 2.1 (continued)

Country & location	Chemicals	Occurrence	Concentration/detection frequency/observation trend	References
Spain (Madrid)	Toluene, p- and m-Xylene, PAHs, and others	Atmosphere (indoor air)	• Maximum levels of PAHs (Naphthalene, Acenaphthene, Phenanthrene, Acenaphthylene) were associated with XAD2	Piñeiro et al. (2021)
Sweden	PFAS	SW	• 3% of total samples from source areas exceed the threshold value of 90 ng/L in PFBA, PFPeA, PFHxA, PFHpA, PFOA, PFNA, PFDA, PFBS, PFHxS, PFOS, and 6:2 FTSA	Gobelius et al. (2018)
Philippines and Thailand Parañaque and Laguna Lake (Philippines); Nonthaburi, central Bangkok and Samut Prakan (Thailand)	PFAS	DW, SW	• 100% detection frequency for PFHpA, PFOA, PFNA, and PFOS in drinking as well as source waters for countries	Guardian et al. (2020)
The USA (Across the country from 19 DWTPs, Chesapeake Bay Watershed, California)	PPCPs	SW, DW	• Median levels were < 10 ng/L, • Source water—Sulfamethoxazole: 12 ng L^{-1}; TCEP: 120 ng L^{-1}; atrazine: 32, 49, and 49 ng L^{-1} in source finished and distribution waters, respectively)	Benotti et al. (2009)
	Hormones and pharmaceuticals		• BPA: 430 ng L^{-1}; SMX: 120 ng L^{-1} meprobamate:164 ng L^{-1} carbamazepine:162 ngL^{-1}, • 1,7 Dimethylxanthine: 416 ng L^{-1}	Bexfield et al. (2019)
	Phytoestrogens and other EDCs		• EDCS detected: sterol cholesterol (88%), the phytoestrogens genistein (79%) and formononetin (55%), the herbicides metolachlor (50%) and atrazine (74%)	Thompson et al. (2021)

(continued)

Table 2.1 (continued)

Country & location	Chemicals	Occurrence	Concentration/detection frequency/observation trend	References
	55 chemicals from (PCPs, cleaners, and other household goods)	Consumer products	• BPA: < 100 µg/g in all samples. Antimicrobials like TCS, *ortho*-phenylphenol, triclocarban, and 1,4-dichlorobenzene, MEAs, glycol ethers, parabens, and phthalates were found in all PPCPs	Dodson et al. (2012)
UK (West Midlands conurbation)	PBDEs and PCBs	Atmosphere	• Home sample—$\sum$PCB: 2530 pg/m^3; $\sum$PBDE:17 pg/m^3 • Indoor air—$\sum$PCB: 588 pg/m3; • Outdoor air—$\sum$PCB: 3820 pg/m^3	Hazrati and Harrad (2006), Jamshidi et al. (2007)

PFAS chemicals can influence our biology by mimicking fatty acids, the components of fat in our bodies, and the foods we consume. They are also considered EDCs because they can interfere with hormone systems (Li et al., 2024). All around the world, different PFASs were found in drinking water; 11 PFASs in the EU exceeded drinking water standards by 3% in Sweden. The highest PFAS in drinking water level was in France with a range of 86–169 ng/L among other countries that were studied, such as Germany, Italy, Spain, Norway, Belgium, and Fore Islands (Domingo and Nadal, 2019). In two developing nations Thailand and the Philippines, 12 PFASs in the range between 7.16 and 54.49 ng/L were found to be present in source and drinking waters (Guardian et al., 2020).

Phthalates and BPA were also identified in Lebanese bottled water. It was established that 59% of the samples of bottled water contained BPA exceeding the detection level set for the study at 0.05 ng/L, BPA content was determined in the range from 0.05 to 1.37 ng/L (Dhaini & Nassif, 2014). In a similar vein, BPA, another EDC that is frequently discovered in drinking water, has been identified in high concentrations in French bottled water and Germany's river waters (up to 0.8 μg/L) (Tursi et al., 2018). BPA levels in river water in Tokyo, Japan, ranged from 2.00 to 230 ng/L, whereas surface water in four Asian nations Japan, China, Korea, and India was tested for BPA and its analouges (Dogra et al., 2024). In China, extremely high concentrations of BPS (19.9–65,600 ng/L) and BPA (75.6–7480 ng/L) were recorded in the Liuxi River stream, followed by BPF (n.d.–474 ng/L) (Huang et al., 2018).

Water pipes have long been coated with coal tar lining; however, the substance also contains PAHs that seep into the supply water, an issue that has been documented previously (Fähnrich et al., 2002). In water flowing through pipes supplied by Stadtwerke Karlsruhe GmbH, Germany, 13 distinct PAHs were found. The majority of the PAHs were found to be adsorbed to the lining's surface and desorbed, but some release was also ascribed to the Gallionella bacteria's iron-oxidizing activity (Maier et al., 2000). In India, the Mithi River, which empties into Mahim Creek near Mumbai, had total PAH concentrations (ΣPAHs) ranging from 157.96 $\pm$ 18.99 μg/L and carcinogenic PAH concentrations (ΣC-PAHs) ranging from 81.31 $\pm$ 9.75 μg/L. The likely sources of these EDCs in the river were the burning of coal, grass, and wood, and the burning of gasoline and diesel (Singare, 2016).

A study conducted in the US between 2006 and 2007 assessed the presence of 51 EDCs in source water, drinking water, and tap water across 19 WWTPs. The findings revealed that atenolol, atrazine, carbamazepine, E1, gemfibrozil, meprobamate, naproxen, phenytoin, sulfamethoxazole (SMX), tris (2-chloroethyl) phosphate (TCEP), and trimethoprim were the most frequently detected contaminants (Benotti et al., 2009). Additionally, a comprehensive nationwide assessment examined EDCs including hormones and pharmaceuticals at 1,091 locations within "Principal Aquifers," which account for approximately 60% of the total volume of groundwater extracted for drinking water supply in the USA. This large-scale systematic evaluation identified the presence of 21 hormones and 103 pharmaceutical compounds, highlighting the widespread occurrence of these contaminants in the nation's water resources (Bexfield et al., 2019).

A study by Singare (2016) found the presence of endocrine-disrupting organochlorine and organophosphorus pesticides in freshwater. The mean level of α- and β-endosulfan, which was 137.75 ng/L, was found to be higher than the U.S. EPA's chronic criteria level of 6.5 ng/L in Mumbai. Aldrin, dieldrin, and endrin were all found to exceed their respective criteria levels set by the U.S. EPA for freshwater aquatic organisms, similarly, chlorpyrifos measuring at 208.77 ng/L exceeded the recommended concentration value set by the Ministry of Environment of British Colombia at < 35 ng/L (Singare, 2016).

In Pakistan, it was discovered that Rb, Mn, Sr, and U naturally contaminated groundwater in the provinces of Sindh and Punjab (Ali et al., 2019). The release of U was thought to be influenced by desorption in oxidized aquifers. A nationwide assessment in Finland found that U and Radon (Rn) were naturally present in groundwater from various regions (Karro & Lahermo, 1999). The trace elements molybdenum (Mo), zinc, cedar, antimony (Sb), iron, bromine (Br), and selenium (Se) were found in significant concentrations in the snow in the Eastern Chinese area as a result of dry deposition, including human sources, according to Gao et al. (2018). Thompson et al. (2021) examined the occurrence of a series of EDCs, such as the phytoestrogens genistein (in 79% of samples) and formononetin (in 55% of samples), in surface water and groundwater samples taken from the three rivers within the Chesapeake Bay Watershed of the USA. One of the largest documented geogenic sources of endocrine-disrupting trace elements with a global distribution is the presence of natural arsenic in groundwater in regions such as Bangladesh, India, Vietnam, Mexico, and Australia. Therefore, Fig. 2.1 illustrates that surface water is exposed to the widest range of EDCs; however, elevated concentrations of specific compounds, such as BPA, in groundwater remain significant. BPA degrades under aerobic conditions in surface water but remains stable in anaerobic soils and anoxic sediments.

2.2.2 *In Water Treatment Plants*

EDCs have been detected globally in influent, effluent, as well as in biosolids and sludge produced by WWTPs, primarily due to the limitations of conventional treatment technologies. As a result, WWTPs are often considered critical hotspots for the accumulation and dissemination of these emerging organic contaminants (Kumar et al., 2023; Silori et al., 2023). Conventional treatment processes exhibit significant constraints in effectively removing EDCs, posing challenges to wastewater management and environmental safety (Kasonga et al., 2021; Silori et al., 2022). However, recent studies indicate that biological treatment methods have demonstrated the capability to remove over 85% of EDCs from raw wastewater, suggesting a promising approach for enhancing the efficiency of WWTPs in mitigating these contaminants.

In Rome, Italy, four estrogens, E3, E2, E1, and EE2, were found in the influents and effluents of six WWTPs (Baronti et al., 2000). The EDCs concentrations in the influent and effluents of a WWTP in Shanghai, China, were found to be 3.56 μg/L

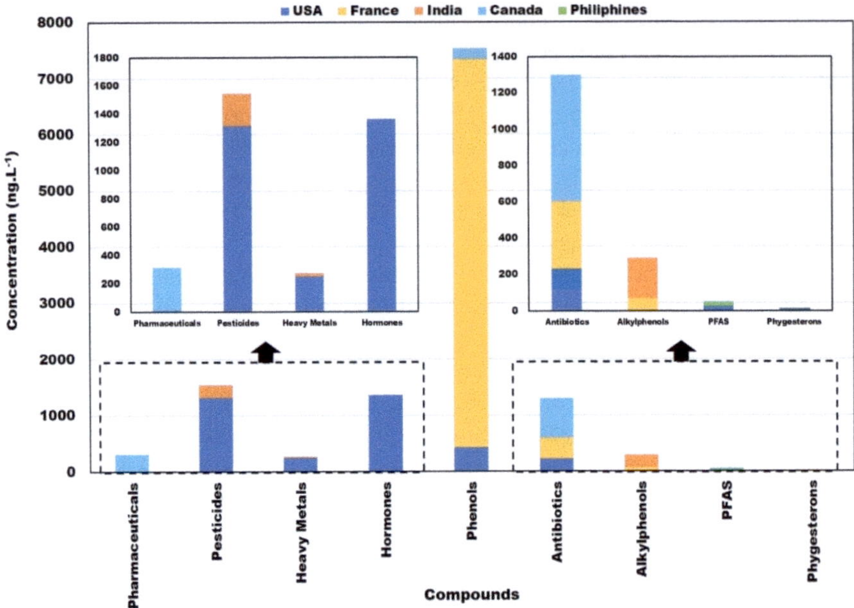

Fig. 2.1 The global prevalence of different classes of endocrine-disrupting compounds (EDCs) that include pesticides, herbicides, BPs, hormones, antibiotics, alkylphenols, isoflavones, and PFAS in four different sample types (groundwater, surface water, source water, &drinking water)

and 1.136 μg/L, respectively (Xu et al., 2016). Furthermore, in a study by Van Zijl et al. (2017), EDCs were found in the WWTP effluents of Cape Town, South Africa, over 93% of the time, with a significant concentration of 5.15 μg/l. In Brazil, the concentration of E1 was observed as 566–242 ng/L, estradiol (E2) at 143–48 ng/L, and ethinylestradiol (EE2) at 421–124 ng/L, while Spain reports estriol (E3) levels at 285–415 ng/L. Germany records the highest bisphenol A (BPA) concentrations at 3640.5–519.5 ng/L, whereas Canada has the highest nonylphenol (NP) levels at 8853–2509 ng/L. In contrast, the lowest EDC concentrations are observed in Australia, with E1 ranging from 8.3 to 10 ng/L and EE2 from < 5.0 to < 0.1 ng/L, while Canada reports the lowest E2 (4 ng/L) and E3 (7–1 ng/L) concentrations. The USA has the lowest NP levels at 510–95 ng/L, while Australia records the lowest BPA concentrations at 648.8–32.0 ng/L. Till now, the highest-ever concentrations of estrone (123.95 μg/l) in the influents of the WWTPs have been observed in the study by Silori et al. (2023).

In a study by Tan et al. (2007) on 5 municipality WWTPs in Australia, 15 EDCs and oestrogen equivalents were measured in grab and passive samples. In grab samples, oestrogen equivalents for influents varied from 108 to 356 ng/L, but for effluents, they were less than 1–14.8 ng/L (Tan et al., 2007). In an Austrian study, a pilot-scale membrane bioreactor (MBR) system operated at different solid retention times (SRT) was compared to several other conventional activated sludge plants (CASPs) also operated at different SRT in treatment efficacy for eight pharmaceuticals, two polycyclic musk fragrances, and nine EDCs (Clara et al., 2005). Some drugs like carbamazepine remained unremoved, while other chemicals like ibuprofen (IBP), BPA, and bezafibrate (BZF) had > 90% removal in both MBR and CASPs; thus, no significant difference with regard to the removal of these compounds was observed between the two. However, MBR resulted in the reduction of particulate matter.

A recent survey from WWTPs in 15 different European countries reveals the presence of 56 different steroids and phenols ranging from a concentration of 25 pg/L (E3) up to 2.4 µg/L (cortisone) in effluents (Finckh et al., 2022). The same study also reported that WWTPs that had better facilities, like ozonation or activated carbon, were found to have comparatively lower EDCs levels in the effluents. The effectiveness of wastewater treatment in reducing EDCs is evident from the variation in their concentrations between influent and effluent streams. While a decrease in contaminant levels is generally expected post-treatment, certain chemicals, such as EDTA, have been observed to exhibit increased concentrations in effluents compared to influents (Barber et al., 2015). This unexpected trend raises concerns about potential internal sources within WWTPs or variations in sampling from different water parcels. These findings highlight the complexities of EDC removal in WWTPs and emphasize the need for further investigation into treatment processes and potential secondary sources contributing to contaminant persistence. A secondary analysis emphasizes the efficiency of wastewater treatment according to EDC concentrations in influents and effluents of Fig. 2.2. Although a reduction in contaminant levels is typically anticipated, some chemicals, like EDTA, were present at higher levels in effluents (Barber et al., 2015). This surprising finding can be explained by internal sources in the treatment system or water sampling variability (Barber et al., 2015).

2.2.3 In Food and Consumer Products

There are hundreds of EDCs in the environment, in food, and in consumer products. They include persistent organic pollutants, pesticides, some heavy metals, preservatives and fragrances, industrial chemicals and their by-products or waste, as well as plant-derived compounds (Singh et al., 2023). These chemicals can enter food through a variety of activities, including contact with packaging materials, processing, agricultural practices (such as using contaminated biosolids as fertilizers), natural production in some food plants, hormones, and antibiotics used in animals (dairy, meat, poultry), among others (Baudry et al., 2021). Sandoval-Insausti et al. (2021) established a correlation between the consumption of conventionally grown fruits and vegetables containing pesticide residues and an increased risk of cancer. Carrots, a staple ingredient in salads, are valued for their high antioxidant content, including alpha- and beta-carotene, lutein, and lycopene (Boadi et al., 2021). However, when cultivated in environments contaminated with PAHs, carrots may pose significant health risks, potentially leading to endocrine disruption in adulthood.

In addition to improving cognitive health, daily eating of table eggs has been shown to prevent age-related macular degeneration (Gopinath et al., 2020). Eggs are

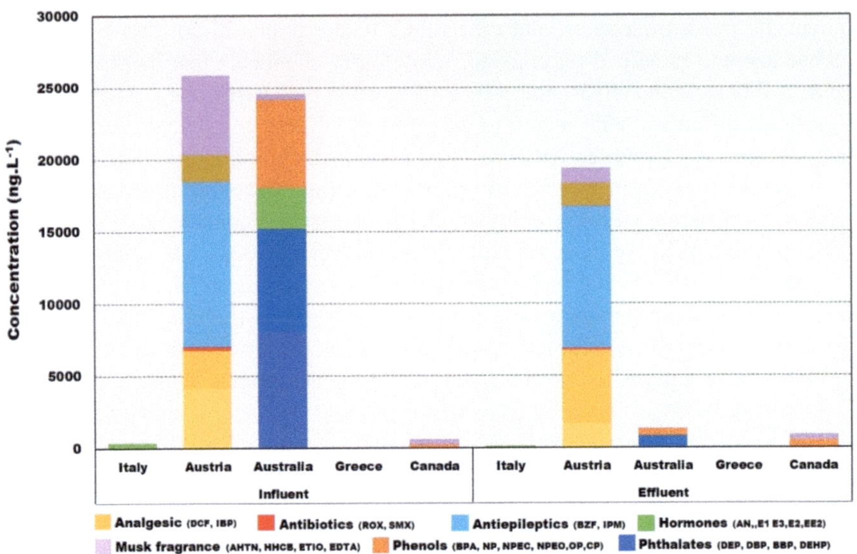

Fig. 2.2 Global concentrations of the different endocrine-disrupting compounds (EDCs) include phthalates, hormones, phenols, analgesics, antiepileptics, and musk fragrances in influent and effluent water of WTTPs

one of the greatest sources of affordable, high-quality protein. Studies have detected endocrine disruptors such as hexachlorobenzene (HCB), polychlorinated naphthalenes (PCNs), polychlorinated dibenzodioxins, and polycyclic aromatic hydrocarbons (PAHs) in eggs from polluted environments (Domingo, 2014; Pajurek et al., 2019). Toxic agents such as polybrominated diphenyl ethers (PBDEs), conventionally used as flame retardants, heavy metals, and estrogenic EDCs in supranormal concentration, have been found in meat products, warranting possible side effects on long-term consumption (Li et al., 2021; Pietron et al., 2019). Bisphenols (BPA, BPF, BPE, BPB, and BPS) have also been reported in various drinks (soda, beer, cola, tea, and energy drinks) from Barcelona, Spain (Gallart-Ayala et al., 2011).

Plant and plant-based foods inherently have bioactive compounds that act as steroid analogs or precursors in humans, which are classified into isoflavones, lignans, and coumestans (Křížová et al., 2019; Tu et al., 2021). Isoflavone's richest sources are legumes of the family papilionoideae, soybeans, chickpea (Cicer arietinum), alfalfa (Medicago sativa), and clover (Gen. Trifolium), containing β-glucosides, malonyl-β-glucoside, acetyl-β-glucoside, and aglucones. Other sources are hops (Humulus lupulus), which is brewed, and dried cherries (Murphy & Hendrich, 2002). Isoflavones have also been found in 56 vegetable forms (Liggins et al., 2000; Murphy & Hendrich, 2002). Isoflavone profiles are greatly transformed by processing. Soymilk manufacture preserves β-glucosides, lowers malonyl-glucosides, and raises aglucones with the action of β-glucosidase (Matsuura and Obata, 1993; Murphy & Hendrich, 2002). Tofu manufacture lowers overall isoflavone content, whereas tempeh manufactured through heat treatment and bacterial fermentation (Brevibacterium epidermidis, Micrococcus luteus, and Microbacterium arborescens) converts glycitein and daidzein to 4,6,7-trihydroxy isoflavone (Klus et al., 1993; Murphy & Hendrich, 2002). Soy flour baking raises β-glucoside levels, whereas frying texturized vegetable protein (TVP) increases acetyl-glucosides (Murphy & Hendrich, 2002). These differences underscore the influence of food processing on the nutritional and functional characteristics of plant foods.

Phthalates and trace elements from the packing materials were found to contaminate brewed coffee (De Toni et al., 2017). Di-isobutyl-phthalate was present in almost all the coffee samples, diethyl-phthalate and di-n-butyl-phthalate were detected in one and three of the four types of capsules, respectively, while Pb and Ni were found in all coffee samples. In today's fast age, fast foods have become popular; however, the high temperatures used for the preparation and processing of these foods lead to the production of a class of EDCs called advanced glycation end products (AGEs, also called Maillard products) (Ravichandran et al., 2019). These EDCs have been linked to diabetes and hasten aging (Schmidt et al., 1992). In a study of human subjects who were fed meat and meat-derived products that were processed under high heat and "dry heating" like broiling, grilling, frying, and roasting, the plasma levels of N-(carboxymethyl) lysine (CML), an AGE compound, were found to be elevated (Ravichandran et al., 2019). Food containers and packaging materials also contain significant sources of organic compounds and EDCs, which ultimately leach into the food (Mezcua et al., 2012).

2.2.4 In Consumer Products

The presence of EDCs in cosmetics and personal care products (PCPs) is well documented, with a wide range of chemicals exhibiting potential endocrine-disruptive properties. Phthalates, parabens, and colophonium are commonly detected in fragrant PCPs such as perfumes, deodorants, and room fresheners (Yang et al., 2015). Fixatives, dyes, and preservatives such as parabens, formaldehyde, glutaraldehyde, derivatives of aromatic amines, metal salts, UV filters, phthalates, solvents, scent compounds, and more are all found in modern cosmetic packaging. While some of these compounds serve as primary active ingredients in product classes like UV filters, others play secondary roles as preservatives or stabilizing formulations. Parabens (p-hydroxybenzoic acid esters) have also been identified in drinking and mineral water sources (Marta-Sanchez et al., 2018; Ripamonti et al., 2018), raising concerns about their persistence and potential exposure routes.

Among antimicrobial agents, triclosan (TCS) is recognized as one of the most relevant EDCs and is widely used in cosmetics, detergents, and other consumer products (Ripamonti et al., 2018). Similarly, UV filters used in sunscreens contain 2-ethylhexyl 4-(dimethylamino) benzoate (Padimate O), octinoxate (ethylhexyl methoxycinnamate), and benzophenones, which exhibit endocrine-disrupting properties and contribute to environmental contamination (Ripamonti et al., 2018). PFAS have also been identified in household products, often used as active ingredients due to their water- and stain-resistant properties. A study by Herzke et al. (2012) conducted in Norway and Sweden analyzed 30 different consumer products and detected PFAS in 27 samples. Notably, perfluorooctane sulfonate (PFOS), a strictly regulated compound in Norway since 2007, exceeded EU regulatory limits in four products, primarily leather and carpet samples. Additionally, fluorotelomer alcohols (FTOHs) were found in carpets, textiles, and waterproofing agents, further contributing to indoor PFAS exposure.

A study by Kotthoff et al. (2015) in Germany revealed high PFAS concentrations in various consumer goods, including ski waxes (max 2000 µg/kg PFOA), leather (max 200 µg/kg PFBA and 120 µg/kg PFBS), outdoor textiles (max 19 µg/m^2 PFOA), and baking papers (max 15 µg/m^2 PFOA), all sourced from local markets. Several products, including carpets, leather, and outdoor materials, exceeded the EU regulatory threshold of 1 µg/m^2 for PFOS, highlighting regulatory challenges in consumer product safety. Expanding on these findings, Bečanová et al. (2016) conducted a large-scale study in the Czech Republic, analyzing 126 product samples, including building materials and car interior components. The study found that 88% of the samples contained at least one PFAS, with PFOS being the most frequently detected compound. The highest concentration of PFAS was found in a textile sample (77.61 mg/kg), and household construction materials, such as composite wood, also exhibited measurable PFAS contamination.

2.2.5 *In the Atmosphere*

Over the past 50 years, new chemicals have increased in the built environment in tandem with the quick development of new consumer goods, furniture, and construction materials. It has been demonstrated that some chemicals found in consumer goods, furnishings, and construction materials are EDCs, meaning they interfere with the function of endogenous hormones (Rudel and Perovich., 2009). Since EDCs have been demonstrated to interfere with normal endocrine signaling in both in vitro and in vivo animal experiments, the biggest worry regarding exposure to these chemicals is the possibility of negative effects on development and reproduction (Colborn et al., 1994).

Higher levels of PCBs in indoor air compared to outdoor air have been caused by the presence of particular products or utilities, such as caulking in older buildings, which is rich in PCBs in many countries like Germany and Switzerland (Rudel & Perovich, 2009). According to a study, PCBs can enter the atmosphere and indoor air through volatilization from sources such as soil (Rayhan et al., 2024; Wee & Grinang., 2024). However, the main cause of the concentration of these compounds in indoor air as opposed to outdoor air is the slow accumulation of these pollutants as a result of venting (Darbre, 2018). Plasticizers such as phthalates have also found applications in spray-based or colloidal consumer goods such as perfumes, hair sprays, and pesticides (as an inert ingredient) (Rudel & Perovich, 2009). Phthalates can readily be leached into the environment including the air from the plastic into which they are incorporated as there is no covalent bonding between the two categories of chemicals, and with aging because the deterioration of the plastics is quicker the leaching process is sped up (Rudel & Perovich, 2009).

Phthalate levels are lower in outdoor and rural areas compared to indoor and urban (Rakkestad et al., 2007). Chain length and branching also come into play, with lighter members such as DEP, DMP, and DBP being more in the air than heavier and less volatile members such as DEHP and benzyl butyl phthalate (BBP) that are more present in dust and other materials (Rudel & Perovich, 2009). Alkyl phenols like pollutants have not been reported at all in the air, although up to 100 ng/m3 of indoor air concentrations were detected. Still, outside levels were relatively low (Rudel & Perovich, 2009). Research on parabens in the household region of Cape Cod, USA, has identified indoor air and house dust to contain three members of the parabens class of chemicals (methyl paraben, ethyl paraben, and butylparaben), with methylparaben being detected most frequently (Rudel & Perovich, 2009).

2.3 Guidelines of EDCs

The lack of international regulation or guidelines regarding EDCs in drinking water is a sign of knowledge gaps and "undervalued perspectives" regarding the possible occurrence of EDCs in different environmental matrices and the health risks. When it comes to the present exposure limits to EDCs, the World Health Organization (WHO) has acknowledged that the drinking water standards are insufficient (WHO, 2012; Wee and Aris, 2019).

2.3.1 U.S. EDCs Guidelines

The Environmental Protection Agency (EPA) is responsible for overseeing the regulations about pesticides and commercial chemicals that are not covered elsewhere. For instance, the United States introduced the Toxic Substances Control Act (TSCA) in 1976, which gave the EPA the power to report, record, test, and limit the use of hazardous chemicals (Wee and Aris, 2017). The corresponding EDC restrictions were then incorporated into the Food Quality Protection Act (FQPA) and the Safe Drinking Water Act (SDWA). To screen EDCs, the Endocrine Disruption Screening Program (EDSP) has developed a two-tier framework: Substances that may interact with the endocrine system (thyroid, androgen, or oestrogen hormone systems) are identified through Tier 1 screening. These chemicals then move on to Tier 2 for additional testing to detect any "adverse endocrine-related effects" that the substance may have and, finally, to establish a quantitative dose-adverse effect relationship. To create the risk assessment report, the Tier 2 results are combined with additional hazard data and exposure evaluation for the particular compound. Lastly, regulatory choices about the chemicals and risk mitigation actions as needed are informed by the risk assessment reports.

The regulation of EDCs in the United States is primarily governed by the Federal Food, Drug, and Cosmetic Act (FD&C Act) of 1938 and the Federal Insecticide, Fungicide, and Rodenticide Act (FIFRA) of 1996. The FD&C Act mandates that manufacturers produce food that is safe, pure, wholesome, and accurately labeled, granting the Food and Drug Administration (FDA) broad regulatory authority over products that fail to meet these standards. However, the Food Additives Amendment of 1958 introduced an exemption for substances generally recognized as safe (GRAS), eliminating the requirement for regulatory oversight if a substance falls under this category (Muncke et al., 2014). Notably, there are no requirements to submit GRAS determinations to the FDA, and a comprehensive review initiated in the 1970s was never completed (Ripamonti et al., 2018). Consequently, since 1982, the FDA has not reconsidered the status of any GRAS substance, leading to the approval of over 10,000 GRAS substances in U.S. food products today (Ripamonti et al., 2018).

A 1997 amendment introduced the concept of food contact substances (FCS) and established regulatory guidance, exempting materials that contribute dietary concentrations below 0.5 µg/kg, except for known or suspected carcinogens (Muncke et al., 2014). Despite these regulations, the FDA has no specific requirements for EDC testing or action following their identification. As a result, EDCs such as nonylphenol, bisphenol A (BPA), tributyltin, triclosan, and several phthalates are legally used in food contact materials, along with polymerization by-products, impurities, and non-intentionally added substances (NIAS), many of which migrate into food (Muncke et al., 2014). Additionally, individual U.S. States enforce their regulations on EDCs in the environment, further complicating the regulatory landscape (Muncke et al., 2014). The existing gaps in federal oversight highlight the need for updated risk assessments and stricter regulatory measures to address the presence of EDCs in consumer products (Kassotis et al., 2020).

2.3.2 EU EDC Regulations

The EU began taking action in 1999 to prioritize compounds for additional review as EDCs, track exposures and effects of EDCs, inform the public about EDCs, and create and verify new testing techniques. To take into consideration their EDC impacts, EU legislation instruments for environmental, health, and consumer protection were gradually modified. In 2018, the EU reiterated its use of the precautionary principle and its goal of reducing total exposure to EDCs, paying special attention to crucial developmental windows.

EDCs are banned in pesticides under the 2009 Plant Protection Products Regulation and the 2012 Biocidal Products Regulation (BPR). Their hazard-based classification aligns with the criteria for carcinogens, mutagens, and reproductive toxicants

(CMRs). It has been documented by Slama et al. (2016) that the BPR has agreed to provide scientific standards for the identification of potential EDCs to the EU by 2013. In 2018, the European Food Safety Authority (EFSA) and the European Chemicals Agency (ECHA) issued guidance on identifying EDCs in pesticides, using data from manufacturers or scientific literature. BPA is subject to stringent regulations and was previously designated as a chemical of very high concern (SVHC) in 2017 (ECHA, 2017).

According to current EU regulations, all chemicals, including possible EDCs, must go through different tests before being put on the market to be sure they won't endanger animals or human health (EC, 2023). These are the steps: 1) Hazard identification: a series of tests certified globally by the OECD are used to determine whether the usage of a target chemical poses a risk of damage to humans or wildlife. 2) Risk assessment: the danger that the planned use of the chemicals poses to the environment and human health is assessed using the data gathered from step 1 and the chemicals' observed environmental levels. The third phase is risk management, where a compound's usage may be limited or further examined if likely potential damage is recognized. The OECD has a five-level methodology for testing and assessment on both mammalian and non-mammalian toxicology, which includes: Level 1 existing data and new or existing non-test information, The OECD website states that Level 2 in vitro assays provide information about specific endocrine mechanisms and pathways (both mammalian and non-mammalian methods); Level 3 in vivo assays provide information about specific endocrine mechanisms and pathways; Level 4 in vivo assays provide information on adverse effects on endocrine relevant endpoints; and Level 5 in vivo assays provide more thorough information on adverse effects on endocrine relevant endpoints over a wider range of the organism's life cycle.

The REACH regulation (2006) governs chemical safety across multiple sectors in the EU, excluding plant protection products, biocides, cosmetics, drugs, and medical device chemicals. Annex XIV mandates that chemicals classified as carcinogenic, mutagenic, reprotoxic (CMRs), persistent, bioaccumulative, and toxic (PBTs), or very persistent and very bioaccumulative (vPvBs) require approval from the European Chemicals Agency (ECHA), regardless of exposure levels. Endocrine-disrupting compounds (EDCs) must also gain approval if deemed of equivalent concern to CMRs, a process that involves extensive evaluation. For chemicals under REACH, hazards must be identified, but usage restrictions follow a risk-based rather than hazard-based approach. As of February 2020, the Substances of Very High Concern (SVHC) list included 205 substances, with 16 flagged for endocrine disruption, subjecting them to stricter regulations. Additionally, 43 substances were placed in Annex XIV, including two recognized EDCs, signaling their phased-out use once viable alternatives become available.

A major deliverable of the European Green Deal, the EU Action Plan: "Towards a Zero Pollution for Air, Water, and Soil" (EC, 2023) was recently endorsed by the European Commission. A watch list of various pollutants is being created based on this, and Member States will have to implement monitoring procedures along the drinking water supply chain and consider if the advisory values are exceeded. The European Commission will add substances that are expected to be present in drinking water to the current list if they are discovered via ongoing screening by member states, including perhaps new EDCs (EC, 2023).

2.3.3 Asian Countries' EDCs Regulations

Asian nations are lacking in terms of their EDC regulatory structure. Japan has been one of the most active countries in Asia. The Ministry of Environment Protection (MEP) launched the Strategic Programs on Environmental Endocrine Disruptors (SPEED) in 1998 as one of the first initiatives in this area (Japanese Environmental Agency, 1998). SPEED focuses on screening EDCs, and authorities have prioritized suspected EDs for examination under this initiative (Japanese Environmental Agency, 1998). The Extended Tasks on Endocrine Disruption (EXTEND), 2010, is a relatively recent initiative run by the Japanese MEP (UNEP, 2017). Its objective is to evaluate the environmental danger that EDCs cause by establishing and putting into practice specific "assessment methodologies" (UNEP, (2017).

The use of BPA in baby bottles was outlawed by the Indian Bureau of Indian Standards (BIS) in 2015 (Mahamuni and Shrinithivihahshini, 2017). India-REACH, also known as the Indian Draft Chemicals (Management and Safety) Rules, is one of the country's current rules; the most recent draft update was released in 2020 (Government of India, 2020). According to the Government of India (2020), India-REACH established the framework for "Notification, Registration, and Restrictions, or Prohibitions, as well as labelling and packaging" requirements for a variety of substances, their mixtures, substances in articles, and intermediates "placed or intended to be placed in the Indian territory."

China's National Environmental Protection Law (2016) included EDCs in the nation's law, defining regulatory action for their monitoring and prohibition. The national policy sought to restrict the use of substances with EDCs properties, with a suggested phasing out entirely by 2017 (UNEP, 2017). In aid of this policy, the Chinese Ministry of Agriculture rolled out the industry standard NY/T 2873–2015, also known as Evaluation Methods of the Endocrine Disruption Effects of Pesticides, officially launched in December 2015 and effective from April 1, 2016. This standard introduces a two-tiered test method in assessing the endocrine-disrupting effects of pesticides, integrating seven in vitro and in vivo testing factors to carry out an integral assessment of probable risks. These regulatory milestones demonstrate China's determination to enhance regulation and reduce the environmental and health hazards of EDCs.

The Republic of Korea has implemented a complex system to address environmental chemicals that come from many sources (ChemSafetyPro, (2015). The following companies deal with industrial chemicals, many of which may be possible EDCs: Occupational Safety and Health Act (OSHA)—Ministry of Employment and Labour; 2) Chemical Control Act (CCA)—Ministry of Environment; 3) Consumer Chemical Products and Biocides Safety Act—Ministry of Environment (K-BPR);

and 4) Act on Registration and Evaluation, of Chemical Substances (K-REACH)—Ministry of Environment (ChemSafetyPro, 2015). The K-REACH (version 2017) restricted substances list contains 12 chemicals, many of which are EDCs such as NP and NPEs. It comprises the registration, notification, hazard evaluation, and risk assessment of new or existing chemicals that are imported or manufactured in Korea (ChemSafetyPro, 2015). The Chemical Safety Act (CCA) was implemented in 2015 and regulates substances that are imported or manufactured in Korea. Information about these substances must be submitted to the Ministry of Environment, and anyone who handles or operates toxic chemicals must "prepare an off-site consequence analysis report that evaluates the impact of a potential chemical accident on the surrounding environment and population" (ChemSafetyPro, 2015).

2.3.4 African Countries' EDCs Regulations

South Africa was the first African country to control EDCs by prohibiting the production, importation, exportation, and sale of infant feeding bottles containing BPA in 2011 (Bornman et al., 2017). A watershed in Africa's strategy for EDC regulation came at the Third International Conference on Chemicals Management (ICCM3) in Nairobi, where the continent aligned itself with more than 100 nations in listing EDCs as an emerging global policy issue (Bornman et al., 2017). This choice recognized the possible negative impacts of EDCs on human and environmental health and strengthened the need to safeguard humans, ecosystems, and their constituents against these toxic substances (Bornman et al., 2017).

In the Strategic Approach to International Chemicals Management (SAICM) event in 2013, where African representatives emphasized key issues about EDCs, further advancements were made. Recognition was that information on exposure to EDCs in human and animal tissues in Africa continues to be limited and that the continent is under greater threat from chemical contamination of agricultural products (Bornman et al., 2017). Considering Africa's singular risks in coping with EDC-associated issues, representatives called for technical support from the United Nations Environment Programme (UNEP) and the World Health Organization (WHO) to prepare recommendations, enhance methods of detecting EDCs, and formulate best practices (Bornman et al., 2017).

In 2015, during the Fourth International Conference on Chemicals Management (ICCM4), almost 100 nations reiterated the necessity of UNEP and WHO support in tackling Africa's EDC issues. Since early-life exposures can result in adult-onset disorders, the meeting concluded that the most crucial exposure period for people, lab animals, and wildlife happens during early development. Therefore, it was stressed that lowering these exposures was a top goal (Bornman et al., 2017). Several regulatory agencies have set drinking water EDC guidelines based on toxicological studies and screening procedures in response to growing concerns. For drinking, bathing, groundwater, and river water, the World Health Organisation (WHO) has established acceptable levels of 4 μg/L and 0.4 μg/L for perfluorooctanoic acid (PFOA) and perfluorooctane sulfonate (PFOS) (WHO, 2017). While GenX chemicals and perfluorobutane sulfonate (PFBS) have advisories of 10 ppt and 2000 ppt, respectively, the U.S. Environmental Protection Agency (EPA) has issued an interim health advisory for PFOA (0.004 ppt) and PFOS (0.02 ppt) in drinking water. A BPA limit of 0.01 μg/L in drinking water has also been established by the European Commission (EC) (EC, 2018). Furthermore, minimum risk values (MRLs) have been set by the Agency for Toxic Substances and Disease Registry (ATSDR) for different EDCs, including trace elements, via oral and inhalation routes (ATSDR, 2023). The increasing awareness of EDC hazards and the necessity of strict regulations to reduce exposure and safeguard the public's health are highlighted by these regulatory actions.

2.3.5　*Oceania EDCs Regulations*

In Australia, the issues of EDCs were initially covered under the Industrial Chemicals (Notification and Assessment) Act of 1989 (ICNA Act), which was administered by the National Industrial Chemicals Notification and Assessment Scheme (NICNAS). Until July 1, 2020, NICNAS was the primary regulator of cosmetics, after which its responsibilities were transferred to the Australian Industrial Chemicals Introduction Scheme (AICIS) (Gabriela et al., 2022). Realizing the possible damage that synthetic chemicals can have on both human health and the environment, AICIS offers regulatory direction on the safe handling of industrial chemicals. The program mainly deals with assessing existing and newly proposed substances (Australian Government, 1989) and setting strong regulations for compounds with endocrine-disrupting activity, such as UV filters found in widely used cosmetic products. Given their potential risks, the endocrine-disrupting action of these chemicals should be scrutinized in public alerts and regulatory analyses.

2.4　Conclusion

The widespread detection of endocrine-disrupting chemicals (EDCs) across water, soil, air, food, and consumer products highlights their persistence and global prevalence. Concentrations ranging from nanograms to micrograms per liter indicate that current wastewater and waste management systems often fail to eliminate these contaminants, allowing continuous recirculation in the environment. The occurrence of compounds such as bisphenols, phthalates, PFAS, and steroid hormones even at low levels raises concern for chronic and combined ("cocktail") exposure effects on humans and wildlife. Regional variations in monitoring and regulation further exacerbate the issue, with many developing regions lacking robust analytical capacity and enforcement frameworks.

A globally coordinated approach is essential to address these challenges through improved detection, risk assessment, and harmonized regulatory standards.

Advancing research on cumulative toxicity and promoting sustainable treatment strategies will be key to mitigating EDC risks. Ultimately, reducing EDC exposure supports global efforts toward SDG 3 (Good Health and Well-being) and SDG 6 (Clean Water and Sanitation), ensuring healthier ecosystems and communities. The regulation of EDCs varies across different regions, with the USA, EU, Asia, Africa, and Oceania adopting distinct approaches based on their regulatory frameworks. Globally, the regulatory environment around EDCs is still scattered, with some areas enacting more stringent laws and others lacking thorough supervision. The USA adopts a risk-based strategy and permits a large number of EDCs in consumer goods, whereas the EU imposes hazard-based prohibitions. Africa still has regulatory and monitoring issues, while Asia and Oceania have taken specific action. To reduce the dangers to human health and the environment posed by EDCs, international rules must be harmonized, risk assessment techniques be improved, and public awareness be raised.

References

Ahmed, F., Bibi, M. H., Ishiga, H., Fukushima, T. & Maruoka, T. (2010). Geochemical study of arsenic and other trace elements in groundwater and sediments of the Old Brahmaputra River plain, Bangladesh. *Environmental Earth Sciences, 60*, 1303–1316. https://doi.org/10.1007/s12 665-009-0270-7.

Ali, W., Aslam, M. W., Feng, C., Junaid, M., Ali, K., Li, S., Chen, Z., Yu, Z., Rasool, A., & Zhang, H. (2019). Unraveling prevalence and public health risks of arsenic, uranium and co-occurring trace metals in groundwater along riverine ecosystem in Sindh and Punjab, Pakistan. *Environmental Geochemistry and Health, 41*, 2223–2238.

ATSDR (2023). Agency for Toxic Substances and Disease Registry Minimal Risk Levels (MRLs). Agency for Toxic Substances and Disease Registry. https://www.atsdr.cdc.gov/mrls/pdfs/ ATSDR%20MRLs%20-%20January%202023%20-%20H.pdf.

Barber, L. B., Loyo-Rosales, J. E., Rice, C. P., Minarik, T. A., & Oskouie, A. K. (2015). Endocrine disrupting alkylphenolic chemicals and other contaminants in wastewater treatment plant effluents, urban streams, and fish in the Great Lakes and Upper Mississippi River Regions. *Science of the Total Environment, 517*, 195–206.

Baronti, C., Curini, R., D'Ascenzo, G., Di Corcia, A., Gentili, A., & Samperi, R. (2000). Monitoring natural and synthetic estrogens at activated sludge sewage treatment plants and in a receiving river water. *Environmental Science & Technology, 34*(24), 5059–5066.

Baudry, J., Rebouillat, P., Allès, B., Cravedi, J. P., Touvier, M., Hercberg, S., Lairon, D., Vidal, R., & Kesse-Guyot, E. (2021). Estimated dietary exposure to pesticide residues based on organic and conventional data in omnivores, pesco-vegetarians, vegetarians and vegans. *Food and Chemical Toxicology, 153*, 112179.

Bečanová, J., Melymuk, L., Vojta, Š., Komprdová, K., & Klánová, J. (2016). Screening for perfluoroalkyl acids in consumer products, building materials and wastes. *Chemosphere, 164*, 322–329.

Benotti, M. J., Trenholm, R. A., Vanderford, B. J., Holady, J. C., Stanford, B. D., & Snyder, S. A. (2009). Pharmaceuticals and endocrine disrupting compounds in US drinking water. *Environmental Science & Technology, 43*(3), 597–603.

Bexfield, L. M., Toccalino, P. L., Belitz, K., Foreman, W. T., & Furlong, E. T. (2019). Hormones and pharmaceuticals in groundwater used as a source of drinking water across the United States. *Environmental Science & Technology, 53*(6), 2950–2960.

Boadi, N. O., Badu, M., Kortei, N. K., Saah, S. A., Annor, B., Mensah, M. B., Okyere, H., & Fiebor, A. (2021). Nutritional composition and antioxidant properties of three varieties of carrot (Daucus carota). *Scientific African, 12*, e00801.

Bornman, M. S., Aneck-Hahn, N. H., De Jager, C., Wagenaar, G. M., Bouwman, H., Barnhoorn, I. E.,... & Heindel, J. J. (2017). Endocrine disruptors and health effects in Africa: a call for action. *Environmental Health Perspectives, 125*(8), 085005.

Chakraborty, P., Bharat, G. K., Gaonkar, O., Mukhopadhyay, M., Chandra, S., Steindal, E. H., & Nizzetto, L. (2022). Endocrine-disrupting chemicals used as common plastic additives: Levels, profiles, and human dietary exposure from the Indian food basket. *Science of The Total Environment, 810*, 152200. https://doi.org/10.1016/j.scitotenv.2021.152200.

ChemSafetyPro, (2015). Overview of Chemical Regulations in China [WWW Document]. http://www.chemsafetypro.com/Topics/China/Overview_of_Chemical_Regulations_in_China.html

Clara, M., Strenn, B., Gans, O., Martinez, E., Kreuzinger, N., & Kroiss, H. (2005). Removal of selected pharmaceuticals, fragrances and endocrine disrupting compounds in a membrane bioreactor and conventional wastewater treatment plants. *Water Research, 39*(19), 4797–4807.

Colborn, T., vom Saal, F. S., & Soto, A. M. (1994). Developmental effects of endocrine-disrupting chemicals in wildlife and humans. *Environmental Impact Assessment Review, 14*(5–6), 469–489.

Colin, A., Bach, C., Rosin, C., Munoz, J. F., & Dauchy, X. (2014). Is drinking water a major route of human exposure to alkylphenol and bisphenol contaminants in France? *Archives of Environmental Contamination and Toxicology, 66*, 86–99. https://doi.org/10.1007/s00244-013-9942-0.

Darbre, P. D. (2018). Overview of air pollution and endocrine disorders. *International Journal of General Medicine, 191–207.

Das, N., Sarma, K. P., Patel, A. K., Deka, J. P., Das, A., Kumar, A., Shea, P. J. & Kumar, M. (2017). Seasonal disparity in the co-occurrence of arsenic and fluoride in the aquifers of the Brahmaputra flood plains, Northeast India. *Environmental Earth Sciences.* https://doi.org/10.1007/s12665-017-6488-x.

Das, N., Das, A., Prasad, K. & Kumar, M. (2018). Provenance, prevalence and health perspective of co-occurrences of arsenic, fluoride and uranium in the aquifers of the Brahmaputra River floodplain. *Chemosphere, 194*, 755–772. https://doi.org/10.1016/j.chemosphere.2017.12.021.

De Toni, L., Tisato, F., Seraglia, R., Roverso, M., Gandin, V., Marzano, C.,... & Foresta, C. (2017). Phthalates and heavy metals as endocrine disruptors in food: a study on pre-packed coffee products. *Toxicology Reports, 4*, 234–239.

Dhaini, H. R., & Nassif, R. M. (2014). Exposure assessment of endocrine disruptors in bottled drinking water of Lebanon. *Environmental Monitoring and Assessment, 186*, 5655–5662.

Dogra, K., Lalwani, D., Dogra, S., Panday, D. P., Raval, N., Trivedi, M., Mora, A., Hernandez, M. S. G., Snyder, S. A., Mahlknecht, J., & Kumar, M. (2024). Indian and global Scenarios of Bisphenols A distribution and its new analogues: Prevalence & probability exceedance. *Journal of Hazardous Materials, 135128.

Domingo, J. L. (2014). Health risks of human exposure to chemical contaminants through egg consumption: A review. *Food Research International, 56*, 159–165.

Domingo, J. L., & Nadal, M. (2019). Human exposure to per-and polyfluoroalkyl substances (PFAS) through drinking water: A review of the recent scientific literature. *Environmental Research, 177*, 108648.

EC (2018). European Commission. Q & A BPA. Eur. Com.

EC (2023). Endocrine disruptors: What is the existing approach in the European Community? Available at: https://ec.europa.eu/environment/chemicals/endocrine/strategy/euapproach_en.htm

ECHA (2017). Member State Committee Unanimously Agrees that Bisphenol A is an Endocrine Disruptor. Available at: https://echa.europa.eu/-/msc-unanimously-agrees-that-bisphenol-a-is-an-endocrine-disruptor#:~:text=MSC%20unanimously%20agrees%20that%20Bisphenol%20A%20is%20an%20endocrine%20disruptor&text=The%20Member%20State%20Committee%20(MSC,serious%20effects%20to%20human%20health

Fähnrich, K. A., Pravda, M., & Guilbault, G. G. (2002). Immunochemical detection of polycyclic aromatic hydrocarbons (PAHs). *Analytical Letters, 35*(8), 1269–1300.

Fekadu, S., Alemayehu, E., Dewil, R., & Van der Bruggen, B. (2019). Pharmaceuticals in freshwater aquatic environments: A comparison of the African and European challenge. *Science of the Total Environment, 654,* 324–337.

Finckh, S., Buchinger, S., Escher, B. I., Hollert, H., König, M., Krauss, M., Leekitratanapisan, W., Schiwy, S., Schlichting, R., Shuliakevich, A., & Brack, W. (2022). Endocrine disrupting chemicals entering European rivers: Occurrence and adverse mixture effects in treated wastewater. *Environment International, 170,* 107608.

Fromme, H., Lahrz, T., Piloty, M., Gebhart, H., Oddoy, A. & Rüden, H. (2004). Occurrence of phthalates and musk fragrances in indoor air and dust from apartments and kindergartens in Berlin (Germany). *Indoor Air, 14,* 188–195. https://doi.org/10.1111/j.1600-0668.2004.00223.x

Fu, P. & Kawamura, K. (2010). Ubiquity of bisphenol A in the atmosphere. *Environmental Pollution, 158,* 3138–3143. https://doi.org/10.1016/j.envpol.2010.06.040.

Furuichi, T., Kannan, K., Giesy, J. P., & Masunaga, S. (2004). Contribution of known endocrine-disrupting substances to the estrogenic activity in Tama River water samples from Japan using instrumental analysis and in vitro reporter gene assay. *Water Research, 38*(20), 4491–4501.

Gabriela, A., Leong, S., Ong, P. S. W., Weinert, D., Hlubucek, J., & Tait, P. W. (2022). Strengthening Australia's Chemical Regulation. *International Journal of Environmental Research and Public Health, 19.* https://doi.org/10.3390/ijerph19116673.

Gallart-Ayala, H., Moyano, E., & Galceran, M. T. (2011). Analysis of bisphenols in soft drinks by on-line solid phase extraction fast liquid chromatography–tandem mass spectrometry. *Analytica Chimica Acta, 683*(2), 227–233.

Ghosh, R., Parde, D., Bhaduri, S., Rajpurohit, P., & Behera, M. (2024). Occurrence, fate, transport, and removal technologies of emerging contaminants: A review on recent advances and future perspectives. *CLEAN–Soil, Air, Water, 52*(12), 2300259.

Glineur, A., Nott, K., Carbonnelle, P., Ronkart, S., & Purcaro, G. (2020). Development and validation of a method for determining estrogenic compounds in surface water at the ultra-trace level required by the EU water framework directive watch list. *Journal of Chromatography A, 1624,* 461242.

Gobelius, L., Hedlund, J., Dürig, W., Tröger, R., Lilja, K., Wiberg, K., & Ahrens, L. (2018). Per- and polyfluoroalkyl substances in Swedish groundwater and surface water: Implications for environmental quality standards and drinking water guidelines. Environmental Science & Technology, *52,* 4340–4349. https://doi.org/10.1021/acs.est.7b05718.

Gopinath, B., Liew, G., Tang, D., Burlutsky, G., Flood, V. M., & Mitchell, P. (2020). Consumption of eggs and the 15-year incidence of age-related macular degeneration. *Clinical Nutrition, 39*(2), 580–584.

Government of India (2020). Indian CMSR (Chemicals Management and Safety Rules). Available at: https://www.cirs-group.com/en/chemicals/india-chemical-management-and-safety-rules-cmsr-reach#:~:text=The%20regulation%20will%20replace%20two,to%20be%20published%20in%202022.

Guardian, M. G. E., Boongaling, E. G., Bernardo-Boongaling, V. R. R., Gamonchuang, J., Boontongto, T., Burakham, R., Arnnok, P., & Aga, D. S. (2020). Prevalence of per-and polyfluoroalkyl substances (PFASs) in drinking and source water from two Asian countries. *Chemosphere, 256,* 127115.

Hazrati, S., & Harrad, S. (2006). Causes of variability in concentrations of polychlorinated biphenyls and polybrominated diphenyl ethers in indoor air. Environmental Science & Technology, *40,* 7584–7589. https://doi.org/10.1021/es0617082.

Herzke, D., Olsson, E., & Posner, S. (2012). Perfluoroalkyl and polyfluoroalkyl substances (PFASs) in consumer products in Norway–A pilot study. *Chemosphere, 88*(8), 980–987.

Huang, C., Wu, L. H., Liu, G. Q., Shi, L., & Guo, Y. (2018). Occurrence and ecological risk assessment of eight endocrine-disrupting chemicals in urban river water and sediments of South China. *Archives of Environmental Contamination and Toxicology, 75,* 224–235.

Jamshidi, A., Hunter, S., Hazrati, S. & Harrad, S. (2007). Concentrations and chiral signatures of polychlorinated biphenyls in outdoor and indoor air and soil in a major U.K. conurbation. *Environmental Science & Technology, 41*, pp.2153–2158.

Karro, E., & Lahermo, P. (1999). Occurrence and chemical characteristics of groundwater in Precambrian bedrock in Finland. In *Special Paper-Geological Survey of Finland* (pp. 85–96).

Kasonga, T. K., Coetzee, M. A., Kamika, I., Ngole-Jeme, V. M., & Momba, M. N. B. (2021). Endocrine-disruptive chemicals as contaminants of emerging concern in wastewater and surface water: A review. *Journal of Environmental Management, 277*, 111485.

Kassotis, C. D., Vandenberg, L. N., Demeneix, B. A., Porta, M., Slama, R., & Trasande, L. (2020). Endocrine-disrupting chemicals: Economic, regulatory, and policy implications. *The Lancet Diabetes & Endocrinology, 8*(8), 719–730.

Kleywegt, S., Pileggi, V., Yang, P., Hao, C., Zhao, X., Rocks, C., Thach, S., Cheung, P., & Whitehead, B. (2011). Pharmaceuticals, hormones and bisphenol A in untreated source and finished drinking water in Ontario, Canada - Occurrence and treatment efficiency. *Science of the Total Environment, 409*, 1481–1488. https://doi.org/10.1016/j.scitotenv.2011.01.010.

Klus, K., Börger-Papendorf, G., & Barz, W. (1993). Formation of 6, 7, 4′-trihydroxyisoflavone (factor 2) from soybean seed isoflavones by bacteria isolated from tempe. *Phytochemistry, 34*(4), 979–981.

Kotthoff, M., Müller, J., Jürling, H., Schlummer, M., & Fiedler, D. (2015). Perfluoroalkyl and polyfluoroalkyl substances in consumer products. *Environmental Science and Pollution Research, 22*(19), 14546–14559.

Křížová, L., Dadáková, K., Kašparovská, J., & Kašparovský, T. (2019). Isoflavones. *Molecules, 24*(6), 1076.

Kumar, M., Das, N., Tripathi, S., Verma, A., Jha, P. K., Bhattacharya, P., & Mahlknecht, J. (2023). Global co-occurrences of multi-(emerging)-contaminants in the hotspots of arsenic polluted groundwater: A pattern of menace. *Current Opinion in Environmental Science & Health, 34*, 100483.

Le Coadou, L., Le Ménach, K., Labadie, P., Dévier, M. H., Pardon, P., Augagneur, S., & Budzinski, H., (2017). Quality survey of natural mineral water and spring water sold in France: Monitoring of hormones, pharmaceuticals, pesticides, perfluoroalkyl substances, phthalates, and alkylphenols at the ultra-trace level. *Science of the Total Environment, 603–604*, 651–662. https://doi.org/10.1016/j.scitotenv.2016.11.174.

Lee, A., Bensaada, S., Lamothe, V., Lacoste, M., & Bennetau-Pelissero, C. (2022). Endocrine disruptors on and in fruits and vegetables: Estimation of the potential exposure of the French population. *Food Chemistry, 373*, 131513.

Lestido-Cardama, A., Petrarca, M., Monteiro, C., Ferreira, R., Marmelo, I., Maulvault, A. L., Anacleto, P., Marques, A., Fernandes, J. O., & Cunha, S. C. (2023). Seasonal occurrence and risk assessment of endocrine-disrupting compounds in Tagus estuary biota (NE Atlantic Ocean coast). *Journal of Hazardous Materials, 444*, 130387.

Li, X., Liu, X., Jia, Z., Wang, T., & Zhang, H. (2021). Screening of estrogenic endocrine-disrupting chemicals in meat products based on the detection of vitellogenin by enzyme-linked immunosorbent assay. *Chemosphere, 263*, 128251.

Li, L., Guo, Y., Ma, S., Wen, H., Li, Y., & Qiao, J. (2024). Association between exposure to per-and perfluoroalkyl substances (PFAS) and reproductive hormones in human: A systematic review and meta-analysis. *Environmental Research, 241*, 117553.

Liggins, J., Grimwood, R., & Bingham, S. A. (2000). Extraction and quantification of lignan phytoestrogens in food and human samples. *Analytical Biochemistry, 287*(1), 102–109.

Lindqvist, N., Tuhkanen, T., & Kronberg, L. (2005). Occurrence of acidic pharmaceuticals in raw and treated sewages and in receiving waters. *Water Research, 39*(11), 2219–2228.

Madsen, S. S., Skovbølling, S., Nielsen, C., & Korsgaard, B. (2004). 17-β Estradiol and 4-nonylphenol delay smolt development and downstream migration in Atlantic salmon. *Salmo Salar. Aquatic Toxicology, 68*(2), 109–120.

Mahamuni, D., & Shrinithivihahshini, N. D. (2017). Need for regulatory policies in India, on the use of bisphenol A in food contact plastic containers. *Current Science, 113*, 861–868.

Maier, M., Maier, D., & Lloyd, B. J. (2000). The role of biofilms in the mobilisation of polycyclic aromatic hydrocarbons (PAHS) from the coal-tar lining of water pipes. *Water Science and Technology, 41*(4–5), 279–285.

Marta-Sanchez, A. V., Caldas, S. S., Schneider, A., Cardoso, S. M. V. S., & Primel, E. G. (2018). Trace analysis of parabens preservatives in drinking water treatment sludge, treated, and mineral water samples. *Environmental Science and Pollution Research, 25*(15), 14460–14470.

Matsuura, M., & Obata, A. (1993). β-Glucosidases from soybeans hydrolyze daidzin and genistin. *Journal of Food Science, 58*(1), 144–147.

Mazur, W. M., Uehara, M., Wähälä, K. & Adlercreutz, H. (2000). Phyto-oestrogen content of berries, and plasma concentrations and urinary excretion of enterolactone after a single strawberry-meal in human subjects. *British Journal of Nutrition, 83*, 381–387.

Metcalfe, C. D., Bayen, S., Desrosiers, M., Muñoz, G., Sauvé, S., & Yargeau, V. (2022). An introduction to the sources, fate, occurrence and effects of endocrine disrupting chemicals released into the environment. *Environmental Research, 207*, 112658.

Mezcua, M., Martínez-Uroz, M. A., Gómez-Ramos, M. M., Gómez, M. J., Navas, J. M., & Fernández-Alba, A. R. (2012). Analysis of synthetic endocrine-disrupting chemicals in food: A review. *Talanta, 100*, 90–106.

Muncke, J., Myers, J. P., Scheringer, M., & Porta, M. (2014). Food packaging and migration of food contact materials: will epidemiologists rise to the neotoxic challenge?. *Journal of Epidemiology and Community Health, 68*(7), 592–594.

Murphy, P. A., & Hendrich, S. (2002). Phytoestrogens in foods.

Nasu, M., Goto, M., Kato, H., Oshima, Y. & Tanaka, H. (2001). Study on endocrine disrupting chemicals in wastewater treatment plants. *Water Science and Technology, 43*, 101–108. https://doi.org/10.2166/wst.2001.0078.

Olatunji, O. S., Fatoki, O. S., Opeolu, B. O., Ximba, B. J., & Chitongo, R. (2017). Determination of selected steroid hormones in some surface water around animal farms in Cape Town using HPLC-DAD. *Environmental Monitoring and Assessment, 189*, 1–10.

Ong, H. T., Samsudin, II., & Soto-Valdez, H. (2022). Migration of endocrine-disrupting chemicals into food from plastic packaging materials: An overview of chemical risk assessment, techniques to monitor migration, and international regulations. *Critical Reviews in Food Science and Nutrition, 62*(4), 957–979.

Pajurek, M., Pietron, W., Maszewski, S., Mikolajczyk, S., & Piskorska-Pliszczynska, J. (2019). Poultry eggs as a source of PCDD/Fs, PCBs, PBDEs and PBDD/Fs. *Chemosphere, 223*, 651–658.

Pietron, W., Pajurek, M., Mikolajczyk, S., Maszewski, S., Warenik-Bany, M., & Piskorska-Pliszczynska, J. (2019). Exposure to PBDEs associated with farm animal meat consumption. *Chemosphere, 224*, 58–64.

Pignotti, E., Farré, M., Barceló, D., & Dinelli, E. (2017). Occurrence and distribution of six selected endocrine disrupting compounds in surface-and groundwaters of the Romagna area (North Italy). *Environmental Science and Pollution Research, 24*, 21153–21167.

Piñeiro, R., Jimenez-Relinque, E., Nevshupa, R. & Castellote, M. (2021). Primary and secondary emissions of VOCs and PAHs in indoor air from a waterproof coal-tar membrane: Diagnosis and remediation. *International Journal of Environmental Research and Public Health, 18*. https://doi.org/10.3390/ijerph182312855.

Pironti, C., Ricciardi, M., Proto, A., Bianco, P. M., Montano, L., & Motta, O. (2021). Endocrine-disrupting compounds: An overview on their occurrence in the aquatic environment and human exposure. *Water, 13*(10), 1347.

Rakkestad, K. E., Dye, C. J., Yttri, K. E., Holme, J. A., Hongslo, J. K., Schwarze, P. E., Becher, R. (2007). Phthalate levels in Norwegian indoor air related to particle size fraction. *Journal of Environmental Monitoring, 9*(12):1419–1425. https://doi.org/10.1039/b709947a. Epub 2007 Sep 7. PMID: 18049782.

Ravichandran, G., Lakshmanan, D. K., Raju, K., Elangovan, A., Nambirajan, G., Devanesan, A. A., & Thilagar, S. (2019). Food advanced glycation end products as potential endocrine disruptors: An emerging threat to contemporary and future generation. *Environment International, 123*, 486–500.

Rayhan, M. R. I., Akbor, M. A., Nahar, A., Chowdhury, N. J., Rahman, M. M., & Saadat, A. H. M. (2024). Exposure of polychlorinated biphenyls via indoor dust particles and their health risks in Dhaka City. *Bangladesh. Journal of Hazardous Materials Advances, 14*, 100421.

Ripamonti, E., Allifranchini, E., Todeschi, S., & Bocchietto, E. (2018). Endocrine disruption by mixtures in topical consumer products. *Cosmetics, 5*(4), 61.

Rudel, R. A., & Perovich, L. J. (2009). Endocrine disrupting chemicals in indoor and outdoor air. *Atmospheric Environment, 43*(1), 170–181.

Saha, S., Narayanan, N., Singh, N., & Gupta, S. (2022). Occurrence of endocrine disrupting chemicals (EDCs) in river water, ground water and agricultural soils of India. *International Journal of Environmental Science and Technology, 19*(11), 11459–11474.

Saito, I., Onuki, A. & Seto, H. (2007). Indoor organophosphate and polybrominated flame retardants in Tokyo. *Indoor Air, 17*, 28–36. https://doi.org/10.1111/j.1600-0668.2006.00442.x.

Sandoval-Insausti, H., Chiu, Y. H., Lee, D. H., Wang, S., Hart, J. E., Mínguez-Alarcón, L., Laden, F., Korat, A. V. A., Birmann, B., Eliassen, A. H., & Willett, W. C. (2021). Intake of fruits and vegetables by pesticide residue status in relation to cancer risk. *Environment International, 156*, 106744.

Schmidt, H. H., Warner, T. D., Ishii, K., Sheng, H., & Murad, F. (1992). Insulin secretion from pancreatic B cells caused by L-arginine-derived nitrogen oxides. *Science, 255*(5045), 721–723.

Sharma, B. M., Bharat, G. K., Chakraborty, P., Martiník, J., Audy, O., Kukučka, P., Přibylová, P., Kukreti, P. K., Sharma, A., Kalina, J., Steindal, E. H., & Nizzetto, L. (2021). A comprehensive assessment of endocrine-disrupting chemicals in an Indian food basket: Levels, dietary intakes, and comparison with European data. *Environmental Pollution, 288*. https://doi.org/10.1016/j.envpol.2021.117750.

Silori, R., Kumar, M., Mahapatra, D. M., Biswas, P., Vellanki, B. P., Mahlknecht, J., Tauseef, S. M., & Barcelo, D. (2023). Prevalence of endocrine disrupting chemicals in the urban wastewater treatment systems of Dehradun, India: Daunting presence of Estrone. *Environmental Research, 235*, 116673.

Silori, R., & Tauseef, S. M. (2022). A review of the occurrence of pharmaceutical compounds as emerging contaminants in treated wastewater and aquatic environments. *Current Pharmaceutical Analysis, 18*(4), 345–379.

Singare, P. U. (2016). Carcinogenic and endocrine-disrupting PAHs in the aquatic ecosystem of India. *Environmental Monitoring and Assessment, 188*(10), 599.

Singh, A., Singh, G., Singh, P., & Mishra, V. K. (2023). Overview of sources, fate, and impact of endocrine disrupting compounds in environment and assessment of their regulatory policies across different continents. *Total Environment Research Themes, 7*, 100071.

Slama, R., Bourguignon, J. P., Demeneix, B., Ivell, R., Panzica, G., Kortenkamp, A., & Zoeller, R. T. (2016). Scientific issues relevant to setting regulatory criteria to identify endocrine-disrupting substances in the European Union. *Environmental Health Perspectives, 124*(10), 1497.

Stasinakis, A. S., Gatidou, G., Mamais, D., Thomaidis, N. S. & Lekkas, T. D. (2008). Occurrence and fate of endocrine disrupters in Greek sewage treatment plants. *Water Research, 42*, 1796–1804.https://doi.org/10.1016/j.watres.2007.11.003.

Tan, B. L., Hawker, D. W., Müller, J. F., Leusch, F. D., Tremblay, L. A., & Chapman, H. F. (2007). Comprehensive study of endocrine disrupting compounds using grab and passive sampling at selected wastewater treatment plants in South East Queensland. *Australia. Environment International, 33*(5), 654–669.

Thompson, T. J., Briggs, M. A., Phillips, P. J., Blazer, V. S., Smalling, K. L., Kolpin, D. W., & Wagner, T. (2021). Groundwater discharges as a source of phytoestrogens and other agriculturally derived contaminants to streams. *Science of the Total Environment, 755*, 142873.

Torres, N. H., Aguiar, M. M., Ferreira, L. F. R., Américo, J. H. P., Machado, Â. M., Cavalcanti, E. B., & Tornisielo, V. L. (2015). Detection of hormones in surface and drinking water in Brazil by LC-ESI-MS/MS and ecotoxicological assessment with Daphnia magna. *Environmental Monitoring and Assessment, 187.* https://doi.org/10.1007/s10661-015-4626-z.

Tu, Y., Yang, Y., Li, Y., & He, C. (2021). Naturally occurring coumestans from plants, their biological activities and therapeutic effects on human diseases. *Pharmacological Research, 169,* 105615.

Tursi, A., Chatzisymeon, E., Chidichimo, F., Beneduci, A., & Chidichimo, G. (2018). Removal of endocrine disrupting chemicals from water: Adsorption of bisphenol-A by biobased hydrophobic functionalized cellulose. *International Journal of Environmental Research and Public Health, 15*(11), 2419.

United Nations Environment Programme (2017). Overview Report I: Worldwide initiatives to identify endocrine disrupting chemicals (EDCs) and potential EDCs. The International Panel on Chemical Pollution (IPCP), 1–40.

Wee, S. Y., & Aris, A. Z. (2017). Ecological risk estimation of organophosphorus pesticides in riverine ecosystems. *Chemosphere, 188,* 575–581.

Wee, S. Y., & Aris, A. Z. (2019). Occurrence and public-perceived risk of endocrine disrupting compounds in drinking water. *NPJ Clean Water, 2*(1), 4.

Wee, S. Y., & Grinang, J. (2024). Endocrine disrupting compounds in the tropical climate of Malaysia: pollution, effects, and management strategies. *Exposure and Health,* 1–18.

WHO (2012). Guidelines for Drinking-Water Quality (4th ed.). Geneva: World Health Organization.

WHO (2017). Regional Office for Europe. Drinking water parameter cooperation project. Support to the revision of Annex I Council Directive 98/83/EC on the quality of water intended for human consumption (Drinking Water Directive). Recommendations. Bonn, 11 September.

Xu, G., Ma, S., Tang, L., Sun, R., Xiang, J., Xu, B.,... & Wu, M. (2016). Occurrence, fate, and risk assessment of selected endocrine disrupting chemicals in wastewater treatment plants and receiving river of Shanghai, China. *Environmental Science and Pollution Research, 23*(24), 25442–25450.

Yang, Q., Zhao, Y., Qiu, X., Zhang, C., Li, R., & Qiao, J. (2015). Association of serum levels of typical organic pollutants with polycystic ovary syndrome (PCOS): a case–control study. *Human Reproduction, 30*(8), 1964–1973.

Zhang, S., Zhang, Q., Darisaw, S., Ehie, O., & Wang, G. (2007). Simultaneous quantification of polycyclic aromatic hydrocarbons (PAHs), polychlorinated biphenyls (PCBs), and pharmaceuticals and personal care products (PPCPs) in Mississippi river water, in New Orleans, Louisiana, USA. *Chemosphere, 66*(6), 1057–1069.

Van Zijl, M. C., Aneck-Hahn, N. H., Swart, P., Hayward, S., Genthe, B., & De Jager, C. (2017). Estrogenic activity, chemical levels and health risk assessment of municipal distribution point water from Pretoria and Cape Town, South Africa. *Chemosphere, 186,* 305–313.

Chapter 3
Exposure, Metabolism, and Toxicity of Endocrine-Disrupting Chemicals

Abstract Endocrine-disrupting chemicals (EDCs) come from a variety of sources, including as industrial discharges, wastewater effluents, and agricultural runoff. They persist because wastewater treatment facilities (WWTPs) are unable to remove them effectively. These pollutants disrupt hormonal signaling by mainly targeting nuclear receptors (NRs), such as oestrogen (ER) and androgen (AR) receptors. These contaminants include pesticides, plastics, heavy metals, and persistent organic pollutants (POPs). In both males and females, exposure to EDC causes reproductive problems, decreased fertility, and developmental abnormalities. Notable effects include ovarian regression, skewed sex ratios, and impaired HPG axis function. The environment and human tissues are home to hazardous substances that can cause cancer, immunotoxicity, and neurodevelopment, such as bisphenols, PCBs, and polychlorinated dioxins. Even with regulatory attempts, exposure continues, especially in poorer countries where rules are still patchy. Improved treatment methods, more stringent laws, and further research are essential to mitigate EDC-related health and environmental impacts.

Keywords Exposure · Metabolism · Toxicity · Pathways · Risk

3.1 Introduction

EDCs were formally defined by the United States Environmental Protection Agency (EPA) in 1996 as exogenous agents that interfere with the synthesis, secretion, transport, binding, action, or elimination of natural hormones responsible for regulating homeostasis, reproduction, development, and behavior. Numerous environmental contaminants, including polychlorinated biphenyls (PCBs) and organochlorine pesticides (OCPs), have demonstrated endocrine-disrupting potential (McKinlay et al., 2008). These chemicals, commonly classified as persistent organic pollutants (POPs), are known for their long-range environmental mobility, resistance to degradation, and bioaccumulative nature, posing significant risks at both ecological and public health levels. Exposure to POPs has been linked to a range of adverse effects, including

M. Kumar et al., *Endocrine Disruptors*,
SpringerBriefs in Molecular Science, https://doi.org/10.1007/978-3-032-15157-5_3

carcinogenicity, neurotoxicity, hepatotoxicity, nephrotoxicity, immunotoxicity, and congenital disorders (Yilmaz et al., 2012).

Human exposure to EDCs occurs predominantly through dietary intake, accounting for over 90% of total chemical exposure, with secondary exposure routes including inhalation and dermal absorption (Cornelis et al., 2012; Gore et al., 2015; Perelló et al., 2012). Mechanistically, many EDCs function by mimicking endogenous hormones and binding to hormone receptors as agonists due to their structural similarity. Others act as antagonists, blocking receptor activity and preventing normal hormonal signaling. In particular, interactions with nuclear hormone receptors often result in altered gene transcription cascades. Beyond receptor-mediated pathways, EDCs may also disrupt cellular functions by targeting membrane-bound receptors, modulating intracellular signaling pathways, or interfering with enzymes responsible for hormone synthesis and metabolism (Warner et al., 2020). Some compounds, such as the surfactant nonylphenol, exert multifaceted effects by simultaneously acting as a weak estrogen receptor agonist and an androgen receptor antagonist, highlighting the complex nature of hormonal disruption. Therefore, this chapter gives an overview of the major routes of exposure of EDCs, which are ingestion, inhalation, and dermal absorption (Bertram et al., 2022). It also gives an overview of how these chemicals are metabolized in the body and points out their toxicological effects on endocrine and other biological systems. The chapter highlights the health hazards of EDC exposure, including reproductive diseases, developmental abnormalities, and hormone-related cancers.

3.2 Pathways and Exposure

Organic pollutants that function as EDCs enter the environment primarily from three sources: runoff from agriculture, commercial and industrial processes, and residential and urban areas (Chakraborty et al., 2024). Both traditional and cutting-edge treatment methods (using membrane bioreactors) in WWTPs are insufficient for the removal of these hazardous substances. It is clear that any contaminants that are not eliminated during wastewater treatment operations wind up in the natural water system and subsequently make their way into food chains (Deblonde et al., 2011). In developing nations, hazardous recycling methods and open burning of solid waste endanger not just the local area where primary emissions take place but also have the potential to affect human health through different exposure pathways. People living close to such dumps may experience seizures, neurological retardation, renal failure, developmental problems, and irreversible harm to their central and peripheral nerve systems. Pesticidal EDCs used in agriculture are susceptible to runoff and volatilization from the soil after application, as established by numerous other studies (Khuman et al., 2020; Rex & Chakraborty, 2022).

Figure 3.1 shows that the main sources of riverine and marine contamination of EDCs are plastic debris and chemicals that have leached from stormwater and rainfall runoff. The final sink for these EDCs in the riverine and marine environment is sedimented in the estuary mixing zone (Gong et al., 2019; Liu et al., 2020). Sewage sludge and manures that contain plant debris also introduce a significant quantity of phytosterols into the environment (Gottschall et al., 2013). Research by Gravert et al. (2021) examined a model explaining how the natural phytosterol β-sitosterol is produced, released, and transformed in agricultural soils. It was proposed that either the sporadic decomposition of plant matter or the addition of biomatter (such as agricultural soil amendments) may cause such phytosterols to function as environmental precursors of steroid hormone synthesis.

Numerous studies have proposed that algae blooms might be another natural source of EDCs (Gong et al., 2019; Jia et al., 2019). Algal blooms were linked to oestrogenic activities in Lake Taihu and other Chinese surface waters (Jia et al., 2019). The potencies of 17 β-estradiol/g dry weight, which ranged from 0.5 to 5.1 ng equivalents, were found to be higher than those found in the sediments of the same study areas (Lei et al., 2015; Lou et al., 2016). Therefore, the geogenic release of endocrine-disrupting trace elements such as As, rubidium (Rb), Mn, strontium (Sr), uranium (U), and others into the environment is another natural mechanism (Ahmed et al., 2010). Two of the most significant sources of environmental contamination caused by EDCs are wastewater treatment plant (WWTP) effluents and the biosolids they produce (Fuzzen et al., 2015; Ngeno et al., 2023).

According to Xu et al. (2009), runoff from a southern California agricultural field irrigated by treated wastewater also contained PPCPs and other EDCs like BPA. This study shows that even in developed nations like the USA may be released into the environment due to inadequate treatment procedures in their WWTPs. International WWTP biosolids have also been shown to contain EDCs (Citulski and Farahbakhsh, 2010; Langdon et al., 2010, 2011). According to the experimental findings, nine of the 18 EDCs were found in the sludge and effluent of the WWTPs in the Chaohu Lake basin, with significant concentrations of DEHP and BPA in the sludge and DEHP in the effluent (Han et al., 2024).

Many other ways that EDCs get into the environment are industrial discharge and airborne emissions. Burning coal releases endocrine-disrupting trace elements such as arsenic (As) (Barik et al., 2025; Yu et al., 2007) and mercury (Hg) (Schug et al., 2016). This is one important airborne pathway. Furthermore, the burning of household garbage, fossil fuels, and internal combustion engine fuels produces polyaromatic hydrocarbons (PAHs) that can alter hormones (Padmanabhan et al., 2021; Sidhu et al., 2005). Industrial wastewater in the Toronto region has been discovered to include nonylphenol (NP), bisphenol A (BPA), 4-tert-octylphenol (OP), and nonylphenol ethoxylates (NPEO). Intermediate sources of these pollutants include the commercial laundry and textile sectors, which use detergents based on NPEO (Lee et al., 2002). According to Kumar et al. (2008), the leather sector has also been found to be a significant source of EDC emissions.

Furthermore, a major source of EDC contamination is firefighting foams, especially aqueous film-forming foams (AFFF), which include per- and polyfluoroalkyl substances (PFASs) that ultimately build up in soil and water systems (Xiao et al., 2017). Environmental pollution is further exacerbated by the direct release of endocrine-disrupting trace elements into the air, water, and soil/sediments by a variety

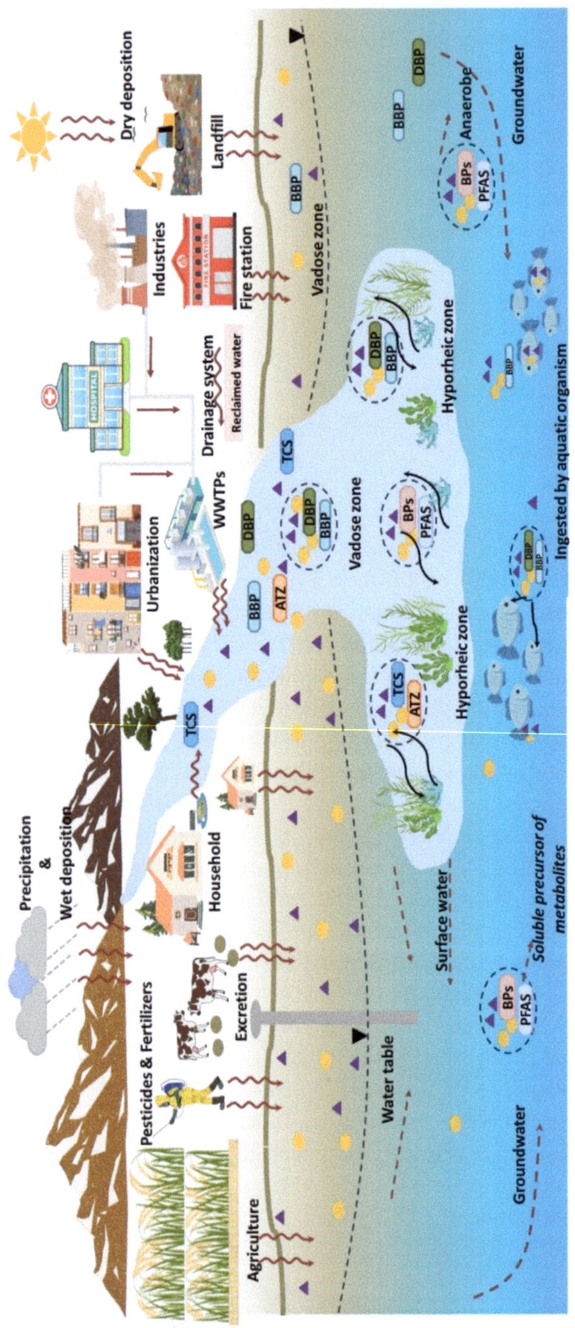

Fig. 3.1 Represents various pathways of exposure to different EDCs in the different environmental matrices: Industrial Sources, agricultural, biosolids and pesticides, WWTPs, PPCPs, consumer products, household stuff, firefighting stations, landfills, hospital waste, and urbanization. All the sources contribute their ways of having EDCs in different water bodies, such as surface water, groundwater, and drinking water. It also represents the ingestion of pollutants by aquatic organisms and human beings

of businesses (Gong et al., 2019). To lessen these substances' negative effects on ecosystems and human health, regulatory action, better industrial waste management, and cutting-edge treatment technologies are desperately needed.

3.3 Metabolism of Endocrine-Disrupting Chemicals

Endocrine disruption refers to the interference of an external substance with the hormonal system, primarily caused by EDCs. Although EDCs can affect any cellular hormonal system, most knowledge is about their effects on nuclear receptor family hormone receptors. A family of 48 structurally similar transcription factors known as nuclear receptors (NRs) governs several essential processes in mammals, including development, homeostasis, reproduction, metabolism, and xenobiotic response (Gronemeyer et al., 2004). ERs, AR, PR, GR, TRs, and mineralocorticoid receptors are important hormone receptors that can be directly targeted by endocrine-disrupting chemicals (EDCs) as either agonists or antagonists.

EDCs activity is also significantly influenced by the aryl hydrocarbon receptor (AhR), a ligand-activated transcription factor involved in xenobiotic metabolism. Organic contaminants such as dioxins, PCBs, PAHs, and endogenous chemicals activate AhR (Poland & Knutson, 1982). According to Rüegg et al. (2007), the transcriptional activation mechanisms of the NR and AhR pathways are comparable and include ligand-induced conformational changes, receptor dimerization (NRs with RXR, AhR with ARNT), and gene transcription. Some NRs, such as TRs, remain DNA-bound, switching between repression and activation upon ligand binding. While direct hormone receptor binding is well-studied (Janosek et al., 2006), recent research highlights cross-talk between NR and AhR signaling as an emerging mechanism of endocrine disruption.

In females, EDCs cause harm by disrupting the hormone physiology that supports the formation and growth of reproductive tissues. An agonistic or antagonistic impact may arise from these exogenous substances interfering with the binding of hormones to their respective receptors, such as the androgen receptor (AR) and the oestrogen receptor (ER) (Sifakis et al., 2017). Dioxins and several OCPs can also bind to AHR, which promotes the induction of CYP1 gene expression. The organochlorine pesticide methoxychlor exhibits oestrogenic action by binding to ERα and ERβ subtypes (Perelló et al., 2012). Therefore, Endosulfan causes ovarian regression, interferes with the hypothalamic-pituitary–gonadal (HPG) axis, and has oestrogenic action in vitro (Warner et al., 2020). By binding to ERα and ERβ receptors, bisphenol A (BPA) has also been demonstrated to have oestrogenic qualities. Furthermore, methoxychlor has an antiandrogenic impact through its interactions with the androgen receptor (AR) (Bertram et al., 2022). Tetrachlorodibenzo-p-dioxin (TCDD), an extremely hazardous environmental contaminant, has also been shown to have antiandrogenic qualities.

The male reproductive system is also affected by EDCs resulting decrease in sperm count, reduced motility, and abnormal sperm morphology. Androgenic or antiandrogenic chemicals contribute significantly to the impairment of male reproductive health development and maintenance (Gore et al., 2015). EDCs have a negative impact on male genital development and reproductive functions through mechanisms that include suppression of aromatase and 5α-reductase (Kogevinas et al., 2001). According to certain theories, environmental contaminants may alter the sex ratio at birth. New data indicate that sperm treated with EDCs like TCDD had a substantially lower Y/X ratio of living spermatozoa than control spermatozoa (Sifakis et al., 2017). These results suggest that a female-biased sex ratio of children at birth might be caused by a decrease in Y sperm viability.

3.4 Toxicity of Endocrine-Disrupting Chemicals

The World Health Organization (WHO) and the Organisation for Economic Co-Operation and Development (OECD) have stated that alterations in the natural hormonal functions (synthesis, secretion, transport, binding, action, and elimination) can result in negative health effects for an intact organism, its offspring, or (sub)populations (Damstra et al., 2002; Fuhrman et al., 2015). Hormonal imbalances, cancer, reproductive dysfunction, and neurological diseases are among the serious health hazards associated with EDCs (Monisha et al., 2023). PFAS and organochlorine pesticides (OCPs) are classified as persistent organic pollutants (POPs) under the Stockholm Convention, and the list is regularly updated to incorporate newly discovered harmful compounds.

EDCs are widely found in consumer goods, air, soil, sediment, water, and food. This causes both acute and long-term exposure, which has harmful biochemical consequences on a variety of species (Monisha et al., 2023). In light of these hazards, toxicological evaluations and studies are essential to comprehending the influence of EDCs. EDCs' toxicity has been studied in recent years using gonadosomatic, in vitro, in silico, embryonic stem cell models, and in vivo methods (Monisha et al., 2023). The initial disruption and cytotoxicity induced by these chemicals can ultimately be manifested as various external and internal symptoms in both humans and wildlife, as depicted in Fig. 3.2. However, there is still inconsistency in the worldwide regulatory frameworks for EDC control and dietary exposure limits. Many developing countries, such as India, lack strict or well-defined rules, which increases population exposure.

The reproductive processes controlled by nuclear receptors (NRs), particularly oestrogen receptors (ERs) and androgen receptors, are significantly impacted by EDCs via the receptor-mediated signal transduction system. More precisely, these substances modify hormone production and breakdown by binding to endocrine receptors (Fig. 3.2). The higher frequency of clear-cell carcinoma in daughters of women administered diethylstilbestrol (DES), a synthetic oestrogen intended to prevent miscarriages, was one of the first documented health crises associated with EDC (Marty et al., 2011). In both male and female children, DES exposure was

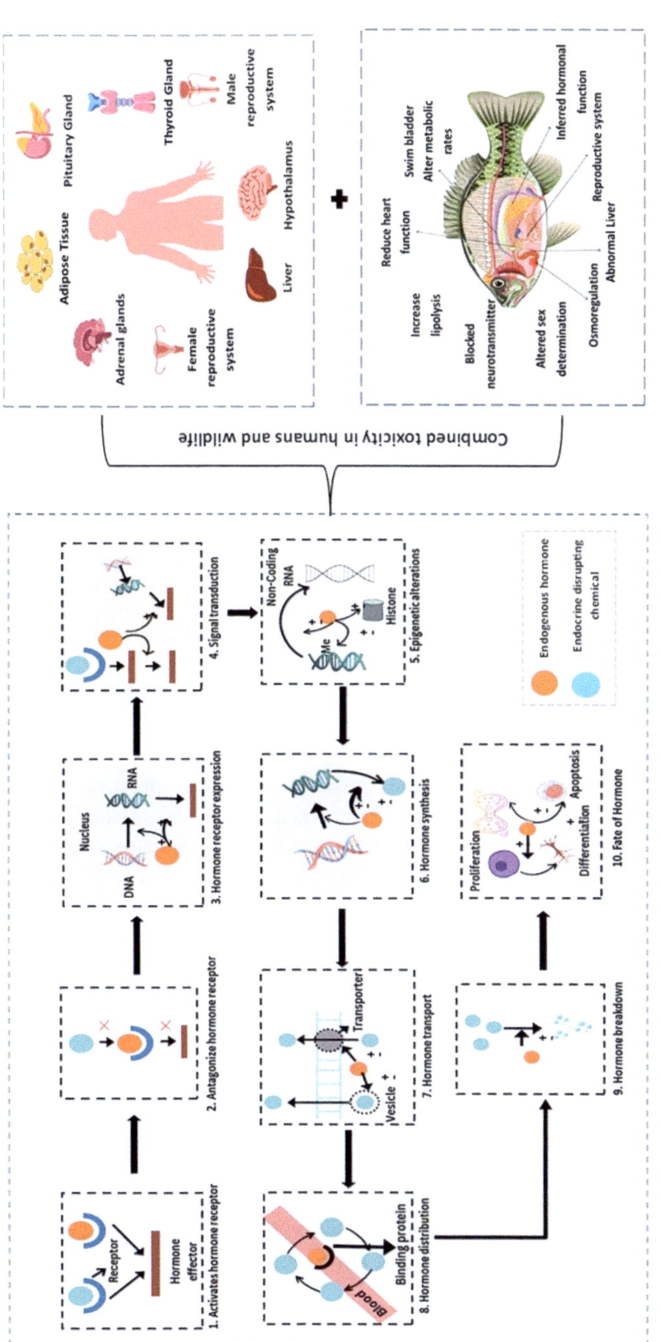

Fig. 3.2 Represents the interaction of EDCs with human and animal physiology (modified and adapted from La Merrill et al., 2020). **1** Activation of hormone receptors, **2** antagonism of hormone receptors, **3** alteration of hormone receptor expression, **4** alteration of signal transduction (including changes in protein or RNA expression, post-translational modifications, and/or ion flux) in hormone-responsive cells, **5** epigenetic modifications in hormone-producing and/or hormone-responsive cells, **6** alteration of hormone synthesis, **7** alteration of hormone transport across the cell membrane, **8** alteration of hormone distribution and circulation levels, **9** alteration of hormone metabolism and/or clearance, and **10** alters the fate of hormone-producing or hormone-responsive cells. Ac, acetyl group; Me, methyl group; EH: endogenous hormone and manifestation of toxic and endocrine-disruptive effects of EDCs in humans and wildlife

also associated with reproductive system anomalies and breast cancer (Giusti et al., 1995). In 1971, the U.S. Food and Drug Administration (FDA) recommended stopping DES because of these negative consequences. Mass poisoning events, such as Yusho disease (1968, Japan) and Yu-Cheng disease (Taiwan), were also caused by exposure to PCBs and polychlorinated dibenzofurans (PCDFs). In these cases, children exposed in utero developed immune dysfunction, cognitive impairments, dermal and ocular lesions, and irregular menstruation as a result of eating contaminated rice (Aoki, 2001).

Female reproductive development, fertility, and the onset of menopause are also affected by EDCs. Although not as well characterized as in males, this collection of anomalies might be an Ovarian Dysgenesis Syndrome, with at least part of the syndrome's origins occurring during fetal development (Fowler et al., 2012). Additionally, EDCs contribute significantly to testicular hypotrophy and polycystic ovarian syndrome (PCOS), as well as a decline in both male and female fertility. Testicular dysgenesis syndrome (TDS) is brought on by EDCs (Zoelle et al., 2012; Campion et al., 2012). Because of harmful environmental factors, impaired spermatogenesis, testicular cancer, undescended testis, and hypopadias may be regarded as signs of a developmental disease. The experiment demonstrated that TDS results from a disruption in gonadal development and embryonal programming throughout fetal life (Ghosh et al., 2022).

PCB exposure causes thymus shrinkage (Yoshimura et al., 1979), altered androgen metabolism (Yoshimura et al., 1985), and alteration of thyroid hormones (Sewall et al., 1995), according to rodent studies. More research on endocrine disruptors has demonstrated how they affect protein kinase modulation, dopamine regulation, and cognitive deficits (Kodavanti & Ward, 1998). A number of EDCs influence reproductive function by acting as agonists or antagonists of the oestrogen receptor (ER) and androgen receptor (AR). At doses as low as 2.5 µg/kg, DES exposure in mice (gestation day 9–16) caused infertility, adenocarcinoma, and sub-fertility (Newbold, 1995). Similarly, mice given dosages of 25–200 mg/kg/day of methoxychlor, an ER and AR agonist, experienced hastened puberty, decreased fertility, and a decrease in sperm count (Gray et al., 1989).

Mostly utilized as ingredients in pesticides, organophosphates are chemicals made via esterification, which combines phosphoric acid and alcohol (Adeyinka et al., 2024). These include diazinon, bensulide, chlorpyrifos, parathion, and malathion, which are mostly used to protect crops and control vectors to limit the development of illnesses conveyed by insects, such as mites and mosquitoes (Qui et al., 2021). Acetylcholinesterase is a crucial enzyme for the release of the neurotransmitter acetylcholine, and these substances impact target species by interfering with its function (Kwong, 2002). Numerous routes of interaction, including direct contact in agricultural fields and ingestion through polluted water, have been recorded because of their extensive usage and abuse, which has led to their detection in a variety of habitats.

In rats, exposure to a coplanar PCB congener has been shown to cause androgen metabolism disruption (Yoshimura et al., 1985) and thymic atrophy (Yoshimura et al., 1979). Other findings include changes in thyroid hormone profiles (Sewall et al., 1995), changes in dopamine levels that lead to poor brain development and cognition (Seegal et al., 1997), and changes in protein kinase C in cultured cerebellar granule cells (Kodavanti & Ward, 1998). The metabolite of DDT, p,p'-DDE (1,1-dichloro-2,2bis(p-chlorophenyl) ethylene) was also found to induce "reduced anogenital distance in male, retention of nipples, as well as altered androgen receptors" by AR antagonism (p,p'-DDE) in laboratory male rats (GD 14–18) at doses of 10 or 100 mg/kg/day (Barlow et al., 1999). The drug finasteride was found to act by 5α-reductase inhibition based on a study on rats (GD 6–20) at doses of 0.0003–100 mg/kg/day, symptoms and effects included reduction of anogenital distance in 4.2% of the male offspring at birth at 0.003 mg/kg/day doses, along with increased rates of hypospadias and retention of the transient nipple in males at doses of $\geq$ 0.3 mg/kg/day (Barlow et al., 1999). The U.S. EPA outlawed the widespread use of DDT in 1972 due to the unacceptably high environmental and health concerns it caused.

According to an in vitro study, BPA, BFP, and BPAF caused human peripheral blood mononuclear cells (PBMCs) to have decreased viability. This was caused by the increased production of reactive oxidative species (ROS, including °OH), which damaged proteins and lipids (Michałowicz et al., 2015). In an in vitro study on human tissue, Cabaton et al. (2009) discovered that BPF and bisphenol f diglycidyl ether (BFDGE) fragmented DNA in HepG2 cells even at non-cytotoxic doses. Research using mutant chicken DT40 cells discovered that BPAP, BPM, and BPP had a greater potential for genotoxicity than BPA (Chen et al., 2016; Lee et al., 2013). Furthermore, BPs toxicity mechanism shows that it was less toxic to the reproductive system and development, had a lower acute toxicity than BPA, and had similar neurotoxicity, immunotoxicity, and endocrine-disruptive effects (Qiu et al., 2019). According to Qiu et al. (2019), BPS acts on the human body through various mechanisms, including oxidative stress, DNA, oestrogenic receptors, and certain protein binding.

3.5 Exposure Risks of Endocrine-Disrupting Chemicals

Polychlorinated dibenzo-pdioxins (PCDDs), and polychlorinated dibenzofurans (PCDFs) are broad terminology that refers to 135 different types of PCDFs and 75 different types of PCDDs. 17 isomers of 2,3,7,8-PCDDs/PCDFs are considered to be hazardous to human health out of these types (Ke et al., 2019). Among the harmful reactions that these stable lipophilic compounds might cause are immunotoxicity, carcinogenicity, reproductive, and developmental problems. A risk evaluation of dioxins and DL-PCBs was carried out in 2001 by the European Commission's Scientific Committee on Food (SCF), which determined a tolerable weekly intake (TWI) of 14 pg TEQ per kg body weight. The European Food Safety Authority (EFSA) CONTAM Panel, however, reevaluated this threshold in 2018 and, as a result of new risk assessments, considerably reduced the TWI to 2 pg TEQ per kg body weight. In humans, the half-life of 2,3,7,8-tetrachlorodibenzo-p-dioxin (TCDD) is noticeably lengthy, spanning from around 7.1 to 11.3 years (Pajurek et al., 2019). In humans,

TCDD has a very long half-life; it has been between 7.1 and 11.3 years. At a WHO meeting in Bilthoven, Netherlands, in December 1990, a tolerated daily intake (TDI) of 10 pg kg^{-1} body weight was established for TCDD (Leeuwen et al., 2000; Patrizi et al., 2018).

Polychlorinated Biphenyls (PCBs) mostly build up in fatty foods that come from animals, especially fish. The Joint FAO/WHO Expert Committee on Food Additives (JECFA) determined a provisional tolerable monthly intake (PTMI) of 70 pg TEQ/kg body weight in 2001, which was in line with the SCF's TWI of 14 pg TEQ/kg body weight per week. The WHO also set a daily threshold for total PCBs at 20 ng/kg body weight (Weber et al., 2018). Therefore, the estimated daily dietary intake of PBDEs, primarily BDE-47, -99, and -153, varies geographically. In the United States, adult exposure is approximately 0.6 ng/kg body weight per day, whereas in Europe, it is higher at 4.0 ng/kg body weight per day (Babalola et al., 2018).

The major way that humans are exposed to BPA is through food. The average amount of total bisphenol in canned foods is considerable, according to different studies, with canned fish having the greatest content (Andújar et al., 2019). In 2018, the allowable amount of BPA from plastics in food contact materials was revised from 0.6 mg/kg to 0.05 mg/kg by the EFSA (Mandel et al., 2019). For adults WHO has provided consumption estimates that range from 0.4 to 4.2 μg/kg body weight per day for adults and from 0.2 to 1.9 μg/kg body weight per day for children over three. According to the U.S. FDA, adults should consume 0.5–1.1 μg/kg body weight of BPA per day, while children aged 12–24 months should consume 0.1–0.3 μg/kg body weight (Geens et al., 2012). In Canada, the U.S. FDA and European Commission prohibited BPA in plastic baby bottles in January 2011 and July 2012, respectively (Liao et al., 2013). It also established the No Observed Adverse Effects Limit (NOAEL) of 5 mg/kg body weight/day for BPA (Mandel et al., 2019). The U.S. EPA and EFSA have established the reference intake level at 50 μg/kg body weight/day, whereas EFSA in 2015 reduced the TDI to 4 from 50 μg/kg body weight/day (Adeyi et al., 2019).

Heavy metals, being naturally present in the Earth's crust, are non-biodegradable, persistent environmental pollutants that accumulate in animals and humans through ingestion, inhalation, and cutaneous absorption (Ameri et al., 2021; Onakpa et al., 2018). The adoption of "potentially toxic elements (PTEs)" as opposed to "heavy metals," particularly for arsenic (As), cadmium (Cd), chromium (Cr), lead (Pb), and mercury (Hg), due to their unparalleled toxicity is favored. U.S. EPA lowered As in water for drinking to 10 ppb from the erstwhile level of 50 μg/L in 2001, while JECFA initially set its PTWI level of 15 μg/kg bw per week in 1989, thereafter revoked in 2010, with EFSA's BMDL$_{01}$ value of 0.3 to 8 μg/kg bw per day limiting a person weighing 70 kg to the intake of just 21 μg/day (Ratnaike et al., 2003; Muñoz et al., 2017).

Hg, anthropogenically released (coal combustion, mining, metallurgy, recycling of electronic waste, incineration of municipal waste) and naturally released (volcanic eruption, weathering of rocks), has a JECFA PTWI of 4 μg/kg bw per week because of its carcinogenicity and chemical resemblance with Zn and As (Tsai et al., 2022; Yutani et al., 2022). Cadmium, the seventh most toxic heavy metal (ATSDR) and

Group 1 human carcinogen (IARC), has a JECFA PTWI of 0.007 mg/kg bw, for which the safe limit of drinking water is established as 0.003 mg/L (WHO) and 0.005 mg/L (EPA). Pb, which has half-lives of 20–30 years in the bone, 40 days in soft tissue, and 35 days in blood, has a JECFA PTWI of 0.025 mg/kg bw, for which the safe limit of drinking water and air is established at 10 μg/L and 0.5 μg/m^3, respectively, whereas 50 μg/L of blood lead level is related to neurodevelopmental impairment, including loss of IQ (Papanikolaou et al., 2005; Stancheva et al., 2014).

3.6 Conclusion

EDCs' persistence, bioaccumulation, and disruption of endocrine processes make them a serious hazard to ecosystems and human health. Their pervasiveness in the air, water, soil, and food supplies leads to a number of negative consequences, such as neurotoxicity, metabolic dysfunction, and reproductive issues. Despite the implementation of regulatory measures to minimize exposure, there are still gaps in the worldwide enforcement of policies, especially in areas with insufficient infrastructure for waste management and treatment. To lessen their long-term effects, laws must be strengthened, wastewater treatment methods must be improved, and scientific research on EDC interactions must be advanced. Effective measures for reducing EDC pollution and preserving ecological and human well-being require a multidisciplinary approach that integrates toxicology, environmental science, and public health activities.

References

Adeyi, A. A., & Babalola, B. A. (2019). Bisphenol-A (BPA) in foods commonly consumed in Southwest Nigeria and its human health risk. *Scientific Reports, 9*(1), 17458.

Adeyinka, A., Muco, E., Regina, A. C., & Pierre, L. (2024). Organophosphates. In *StatPearls*. Treasure Island (FL), StatPearls Publishing. https://pubmed.ncbi.nlm.nih.gov/29763035.

Ahmed, F., Bibi, M. H., Ishiga, H., Fukushima, T., & Maruoka, T. (2010). Geochemical study of arsenic and other trace elements in groundwater and sediments of the Old Brahmaputra River Plain, Bangladesh. *Environmental Earth Sciences, 60*, 1303–1316.

Ameri, Z., Hoodaji, M., Rajaie, M., & Ataabadi, M. (2021). Optimizing modified rice bran for treating aqueous solutions polluted by Cr (VI) ions: Isotherm and kinetics analyses. *Quality Assurance and Safety of Crops & Foods, 13*(SP1), 1–11.

Andújar, N., Gálvez-Ontiveros, Y., Zafra-Gómez, A., Rodrigo, L., Álvarez-Cubero, M. J., Aguilera, M., Monteagudo, C., & Rivas, A. (2019). Bisphenol A analogues in food and their hormonal and obesogenic effects: A review. *Nutrients, 11*(9), 2136.

Aoki, Y. (2001). Polychlorinated biphenyls, polychloronated dibenzo-p-dioxins, and polychlorinated dibenzofurans as endocrine disrupters—what we have learned from Yusho disease. *Environmental Research, 86*(1), 2–11.

Babalola, B. A., & Adeyi, A. A. (2018). Levels, dietary intake and risk of polybrominated diphenyl ethers (PBDEs) in foods commonly consumed in Nigeria. *Food Chemistry, 265*, 78–84.

Barlow, S. et al. (1999). Teratology Society public affairs committee position paper: Developmental toxicity of endocrine disrupters to humans. *Teratology, 60*(6), 365–375. https://doi.org/10.1002/(SICI)1096-9926(199912)60:63.0.CO;2-6.

Barik, D., Rakhi Mol, K. M., Anand, G., Nandamol, P. S., Das, D., & Porel, M. (2025). Environmental pollutants such as endocrine disruptors/pesticides/reactive dyes and inorganic toxic compounds metals, radionuclides, and metalloids and their impact on the ecosystem. In *Biotechnology for Environmental Sustainability* (pp. 391–442). Singapore: Springer Nature Singapore.

Bertram, M. G., Gore, A. C., Tyler, C. R., & Brodin, T. (2022). Endocrine-disrupting chemicals. *Current Biology, 32*(13), R727–R730.

Cabaton, N., Dumont, C., Severin, I., Perdu, E., Zalko, D., Cherkaoui-Malki, M., & Chagnon, M. C. (2009). Genotoxic and endocrine activities of bis (hydroxyphenyl) methane (bisphenol F) and its derivatives in the HepG2 cell line. *Toxicology, 255*(1–2), 15–24.

Campion, S., Catlin, N., Heger, N., McDonnell, E. V., Pacheco, S. E., Saffarini, C., Sandrof, M. A., & Boekelheide, K. (2012). Male reprotoxicity and endocrine disruption. In *Molecular, Clinical and Environmental Toxicology: Volume 3: Environmental Toxicology* (pp.315–360).

Chakraborty, P., Chandra, S., Pavithra, K., Mukhopadhyay, M., Singh, D., Bera, M., & Sharma, B. M. (2024). Exposure pathway and risk assessment of endocrine-disrupting chemicals. In *Endocrine-Disrupting Chemicals* (pp. 251–277). Elsevier.

Chen, D., Kannan, K., Tan, H., Zheng, Z., Feng, Y. L., Wu, Y., & Widelka, M. (2016). Bisphenol analogues other than BPA: Environmental occurrence, human exposure, and toxicity a review. *Environmental Science & Technology, 50*(11), 5438–5453.

Citulski, J. A., & Farahbakhsh, K., (2010). Fate of endocrine-active compounds during municipal biosolids treatment: A review. *Environmental Science & Technology, 44*, 8367–8376. https://doi.org/10.1021/es102403y.

Cornelis, C., D'Hollander, W., Roosens, L., Covaci, A., Smolders, R., Van Den Heuvel, R., Govarts, E., Van Campenhout, K., Reynders, H., & Bervoets, L. (2012). First assessment of population exposure to perfluorinated compounds in Flanders. *Belgium. Chemosphere, 86*(3), 308–314.

Damstra, T., Barlow, S., Bergman, A., Kavlock, R., & Van Der Kraak, G. E. (2002). *International programme on chemical safety global assessment: The state-of-the-science of endocrine disruptors*. World Health Organization.

Deblonde, T., Cossu-Leguille, C., & Hartemann, P. (2011). Emerging pollutants in wastewater: A review of the literature. *International Journal of Hygiene and Environmental Health, 214*(6), 442–448.

Fowler, P. A., Bellingham, M., Sinclair, K. D., Evans, N. P., Pocar, P., Fischer, B., Schaedlich, K., Schmidt, J. S., Amezaga, M. R., Bhattacharya, S., & Rhind, S. M. (2012). Impact of endocrine-disrupting compounds (EDCs) on female reproductive health. *Molecular and Cellular Endocrinology, 355*(2), 231–239.

Fuhrman, V. F., Tal, A., & Arnon, S. (2015). Why endocrine disrupting chemicals (EDCs) challenge traditional risk assessment and how to respond. *Journal of Hazardous Materials, 286*, 589–611.

Fuzzen, M. L., Bennett, C. J., Tetreault, G. R., McMaster, M. E., & Servos, M. R. (2015). Severe intersex is predictive of poor fertilization success in populations of rainbow darter (Etheostoma caeruleum). *Aquatic Toxicology, 160*, 106–116.

Geens, T., Aerts, D., Berthot, C., Bourguignon, J. P., Goeyens, L., Lecomte, P., Maghuin-Rogister, G., Pironnet, A. M., Pussemier, L., Scippo, M. L., & Van Loco, J. (2012). A review of dietary and non-dietary exposure to bisphenol-A. *Food and Chemical Toxicology, 50*(10), 3725–3740.

Ghosh, A., Tripathy, A., & Ghosh, D. (2022). Impact of endocrine disrupting chemicals (EDCs) on reproductive health of human. *Proceedings of the Zoological Society, 75*(1), 16–30. New Delhi: Springer India.

Giusti, R. M., Iwamoto, K., & Hatch, E. E. (1995). Diethylstilbestrol revisited: A review of the long-term health effects. *Annals of Internal Medicine, 122*(10), 778–788.

Gong, J., Ran, Y., Zhang, D., Chen, D., Li, H., & Huang, Y. (2019). Vertical profiles and distributions of aqueous endocrine-disrupting chemicals in different matrices from the Pearl River Delta and the influence of environmental factors. *Environmental Pollution, 246*, 328–335.

Gore, A. C., Chappell, V. A., Fenton, S. E., Flaws, J. A., Nadal, A., Prins, G. S., Toppari, J., & Zoeller, R. T. (2015). EDC-2: The endocrine society's second scientific statement on endocrine-disrupting chemicals. *Endocrine Reviews, 36*(6), E1–E150.

Gottschall, N., Topp, E., Edwards, M., Payne, M., Kleywegt, S., Russell, P., & Lapen, D. R. (2013). Hormones, sterols, and fecal indicator bacteria in groundwater, soil, and subsurface drainage following a high single application of municipal biosolids to a field. *Chemosphere, 91*(3), 275–286.

Gravert, T. K. O., Fauser, P., Olsen, P., & Hansen, M. (2021). In situ formation of environmental endocrine disruptors from phytosterol degradation: A temporal model for agricultural soils. *Environmental Science: Processes and Impacts, 23*, 855–866. https://doi.org/10.1039/d1em00 027f.

Gray, L. E., Ostby, J., Ferrell, J., Rehnberg, G., Biology, D. R., Sections, G. B., & Carolina, N. (1989). A dose-response analysis of methoxychlor-induced alterations reproductive development and function in the rat. *Fundamental and Applied Toxicology, 12*, 92–108.

Gronemeyer, H., Gustafsson, J. Å., & Laudet, V. (2004). Principles for modulation of the nuclear receptor superfamily. *Nature Reviews Drug Discovery, 3*(11), 950–964.

Han, Y., Qi, C., Niu, Z., Li, N., & Tang, J. (2024). Occurrence and risk assessment of endocrine-disrupting chemicals in wastewater treatment plants in the Chaohu Lake Basin. *Frontiers in Environmental Science, 12*, 1409011.

Janošek, J., Hilscherová, K., Bláha, L., & Holoubek, I. (2006). Environmental xenobiotics and nuclear receptors—Interactions, effects and in vitro assessment. *Toxicology in Vitro, 20*(1), 18–37.

Jia, Y., Chen, Q., Crawford, S. E., Song, L., Chen, W., Hammers-Wirtz, M., Strauss, T., Seiler, T. B., Schäffer, A., & Hollert, H. (2019). Cyanobacterial blooms act as sink and source of endocrine disruptors in the third largest freshwater lake in China. *Environmental Pollution, 245*, 408–418.

Ke, X., Qi, Y., Bao, Q., & Zhang, H. (2019). Concentrations, sources, and TEQ of PCDD/Fs in sediments from the liaohe river protected areas. *Archives of Environmental Contamination and Toxicology, 76*, 171–177.

Khuman, S. N., Vinod, P. G., Bharat, G., Kumar, Y. M., & Chakraborty, P. (2020). Spatial distribution and compositional profiles of organochlorine pesticides in the surface soil from the agricultural, coastal and backwater transects along the south-west coast of India. *Chemosphere, 254*, 126699.

Kodavanti, P. R., & Ward, T. R. (1998). Interactive effects of environmentally relevant polychlorinated biphenyls and dioxins on [3H] phorbol ester binding in rat cerebellar granule cells. *Environmental Health Perspectives, 106*(8), 479–486.

Kogevinas, M. (2001). Human health effects of dioxins: Cancer, reproductive and endocrine system effects. *APMIS, 109*(S103), S223–S232.

Kumar, V., Majumdar, C., & Roy, P. (2008). Effects of endocrine disrupting chemicals from leather industry effluents on male reproductive system. *The Journal of Steroid Biochemistry and Molecular Biology, 111*(3–5), 208–216.

Kwong, T. C. (2002). Organophosphate pesticides: Biochemistry and clinical toxicology. *Therapeutic Drug Monitoring, 24*(1), 144–149.

Langdon, K. A., Warne, M. S. J., Smernik, R. J., Shareef, A., & Kookana, R. S. (2011). Selected personal care products and endocrine disruptors in biosolids: An Australia-wide survey. *Science of The Total Environment, 409*, 1075–1081. https://doi.org/10.1016/j.scitotenv.2010.12.013.

Langdon, K. A., Warne, M. S. T. J., & Kookana, R. S. (2010). Aquatic hazard assessment for pharmaceuticals, personal care products, and endocrine-disrupting compounds from biosolids-amended land. *Integrated Environmental Assessment and Management, 6*(4), 663–676. https:// doi.org/10.1002/ieam.74.

Lee, H. B., Peart, T. E., Gris, G., & Chan, J. (2002). Endocrine-disrupting chemicals in industrial wastewater samples in Toronto, Ontario. *Water Quality Research Journal, 37*(2), 459–472.

Lee, S., Liu, X., Takeda, S., & Choi, K. (2013). Genotoxic potentials and related mechanisms of bisphenol A and other bisphenol compounds: A comparison study employing chicken DT40 cells. *Chemosphere, 93*(2), 434–440.

Lei, B., Kang, J., Wang, X., Liu, Q., Yu, Z., Zeng, X., & Fu, J. (2015). The toxicity of sediments from Taihu Lake evaluated by several in vitro bioassays. *Environmental Science and Pollution Research, 22,* 3419–3430.

Liao, C., & Kannan, K. (2013). Concentrations and profiles of bisphenol A and other bisphenol analogues in foodstuffs from the United States and their implications for human exposure. *Journal of Agricultural and Food Chemistry, 61*(19), 4655–4662.

Liu, Y., Su, W., Zhu, Y., Xiao, L., & Hu, T. (2020). Endocrine disrupting compounds in the middle and lower reaches of the Lhasa River Basin: Occurrence, distribution, and risk assessment. *Science of the Total Environment, 727,* 138694.

Lou, S., Lei, B., Feng, C., Xu, J., Peng, W., & Wang, Y. (2016). In vitro toxicity assessment of sediment samples from Huangpu River and Suzhou River, Shanghai, China. *Environmental Science and Pollution Research, 23,* 15183–15192.

Mandel, N. D., Gamboa-Loira, B., Cebrián, M. E., Mérida-Ortega, Á., & López-Carrillo, L. (2019). Challenges to regulate products containing bisphenol A: Implications for policy. *Salud Pública De México, 61*(5), 692–697.

Marty, M. S., Carney, E. W., & Rowlands, J. C. (2011). Endocrine disruption: historical perspectives and its impact on the future of toxicology testing. *Toxicological Sciences, 120*(suppl_1), S93–S108.

McKinlay, R., Plant, J. A., Bell, J. N. B., & Voulvoulis, N. (2008). Endocrine disrupting pesticides: Implications for risk assessment. *Environment International, 34*(2), 168–183.

Michałowicz, J., Mokra, K., & Bąk, A. (2015). Bisphenol A and its analogs induce morphological and biochemical alterations in human peripheral blood mononuclear cells (in vitro study). *Toxicology in Vitro, 29*(7), 1464–1472.

Monisha, R. S., Mani, R. L., Sivaprakash, B., Rajamohan, N., & Vo, D. V. N. (2023). Remediation and toxicity of endocrine disruptors: A review. *Environmental Chemistry Letters, 21*(2), 1117–1139.

Muñoz, O., Zamorano, P., Garcia, O., & Bastías, J. M. (2017). Arsenic, cadmium, mercury, sodium, and potassium concentrations in common foods and estimated daily intake of the population in Valdivia (Chile) using a total diet study. *Food and Chemical Toxicology, 109,* 1125–1134.

Newbold, R. (1995). Cellular and molecular effects of developmental exposure to diethylstilbestrol: Implications for other environmental estrogens. *Environmental Health Perspectives, 103*(suppl 7), 83–87.

Ngeno, E., Ongulu, R., Orata, F., Matovu, H., Shikuku, V., Onchiri, R., Mayaka, A., Majanga, E., Getenga, Z., Gichumbi, J., & Ssebugere, P. (2023). Endocrine disrupting chemicals in wastewater treatment plants in Kenya, East Africa: Concentrations, removal efficiency, mass loading rates and ecological impacts. *Environmental Research, 237,* 117076.

Onakpa, M. M., Njan, A. A., & Kalu, O. C. (2018). A review of heavy metal contamination of food crops in Nigeria. *Annals of Global Health, 84*(3), 488.

Padmanabhan, V., Song, W., & Puttabyatappa, M. (2021). Praegnatio perturbatio—Impact of endocrine-disrupting chemicals. *Endocrine Reviews, 42*(3), 295–353.

Pajurek, M., Pietron, W., Maszewski, S., Mikolajczyk, S., & Piskorska-Pliszczynska, J. (2019). Poultry eggs as a source of PCDD/Fs, PCBs, PBDEs and PBDD/Fs. *Chemosphere, 223,* 651–658.

Papanikolaou, N. C., Hatzidaki, E. G., Belivanis, S., Tzanakakis, G. N., & Tsatsakis, A. M. (2005). Lead toxicity update. A brief review. *Medical science monitor, 11*(10), RA329.

Patrizi, B., & Siciliani de Cumis, M. (2018). TCDD toxicity mediated by epigenetic mechanisms. *International Journal of Molecular Sciences, 19*(12), 4101.

Perelló, G., Gómez-Catalán, J., Castell, V., Llobet, J. M., & Domingo, J. L. (2012). Assessment of the temporal trend of the dietary exposure to PCDD/Fs and PCBs in Catalonia, over Spain: Health risks. *Food and Chemical Toxicology, 50*(2), 399–408.

Poland, A., & Knutson, J. C. (1982). 2, 3, 7, 8-tetrachlorodibenzo-p-dioxin and related halogenated aromatic hydrocarbons: Examination of the mechanism of toxicity. *Annual Review of Pharmacology and Toxicology, 22*, 517–554.

Qiu, W., Zhan, H., Hu, J., Zhang, T., Xu, H., Wong, M., Xu, B., & Zheng, C. (2019). The occurrence, potential toxicity, and toxicity mechanism of bisphenol S, a substitute of bisphenol A: A critical review of recent progress. *Ecotoxicology and Environmental Safety, 173*, 192–202.

Qiu, J., Wheeler, S. S., Reed, M., Goodman, G. W., Xiong, Y., Sy, N. D., Ouyang, G., & Gan, J. (2021). When vector control and organic farming intersect: Pesticide residues on rice plants from aerial mosquito sprays. *Science of the Total Environment, 773*, 144708.

Ratnaike, R. N. (2003). Acute and chronic arsenic toxicity. *Postgraduate Medical Journal, 79*(933), 391–396.

Rex, K. R., & Chakraborty, P. (2022). Legacy and new chlorinated persistent organic pollutants in the rivers of south India: Occurrences, sources, variations before and after the outbreak of the COVID-19 pandemic. *Journal of Hazardous Materials, 437*, 129262.

Rüegg, J., Swedenborg, E., Wahlstrom, D., Escande, A., Balaguer, P., Pettersson, K., & Pongratz, I. (2007). The transcription factor ARNT functions as an estrogen receptor beta selective co-activator, and its recruitment to alternative pathways mediates anti-estrogenic effects of TCDD. *Molecular Endocrinology, 22*, 304–316.

Schug, T. T., Johnson, A. F., Birnbaum, L. S., Colborn, T., Guillette, L. J., Jr., Crews, D. P., Collins, T., Soto, A. M., Vom Saal, F. S., McLachlan, J. A., & Sonnenschein, C. (2016). Minireview: Endocrine disruptors: Past lessons and future directions. *Molecular Endocrinology, 30*(8), 833–847.

Seegal, R. F., Brosch, K. O., & Okoniewski, R. J. (1997). Effects of in utero and lactational exposure of the laboratory rat to 2,4,2',4'- and 3,4,3',4'-tetrachlorobiphenyl on dopamine function. *Toxicology and Applied Pharmacology, 146*, 95–103. https://doi.org/10.1006/taap.1997.8226.

Sewall, C., Flagler, N., Vandenheuvel, J. P., Clark, G. C., Tritscher, A. M., Maronpot, R. M., & Lucier, G. W. (1995). Alterations in thyroid function in female Sprague-Dawley rats following chronic treatment with 2, 3, 7, 8-tetrachlorodibenzo-p-dioxin. *Toxicology and Applied Pharmacology, 132*(2), 237–244.

Sidhu, S., Gullett, B., Striebich, R., Klosterman, J., Contreras, J., & DeVito, M. (2005). Endocrine disrupting chemical emissions from combustion sources: Diesel particulate emissions and domestic waste open burn emissions. *Atmospheric Environment, 39*(5), 801–811.

Sifakis, S., Androutsopoulos, V. P., Tsatsakis, A. M., & Spandidos, D. A. (2017). Human exposure to endocrine disrupting chemicals: Effects on the male and female reproductive systems. *Environmental Toxicology and Pharmacology, 51*, 56–70.

Stancheva, M., Makedonski, L., & Peycheva, K. (2014). Determination of heavy metal concentrations of most consumed fish species from Bulgarian Black Sea coast. *Bulgarian Chemical Communications, 46*(1), 195–203.

Tsai, W. T. (2022). Multimedia pollution prevention of mercury-containing waste and articles: Case study in Taiwan. *Sustainability, 14*(3), 1557.

van Leeuwen, F. R., Feeley, M., Schrenk, D., Larsen, J. C., Farland, W., & Younes, M. (2000). Dioxins: WHO's tolerable daily intake (TDI) revisited. *Chemosphere, 40*(9–11), 1095–1101.

Warner, G. R., Mourikes, V. E., Neff, A. M., Brehm, E., & Flaws, J. A. (2020). Mechanisms of action of agrochemicals acting as endocrine disrupting chemicals. *Molecular and Cellular Endocrinology, 502*, 110680.

Weber, R., Herold, C., Hollert, H., Kamphues, J., Ungemach, L., Blepp, M., & Ballschmiter, K. (2018). Life cycle of PCBs and contamination of the environment and of food products from animal origin. *Environmental Science and Pollution Research, 25*, 16325–16343.

Xiao, X., Ulrich, B. A., Chen, B., & Higgins, C. P. (2017). Sorption of poly-and perfluoroalkyl substances (PFASs) relevant to aqueous film-forming foam (AFFF)-impacted groundwater by biochars and activated carbon. *Environmental Science & Technology, 51*(11), 6342–6351.

Xu, J., Wu, L., Chen, W., Jiang, P. & Chang, A. C. S. (2009). Pharmaceuticals and personal care products (PPCPs), and endocrine disrupting compounds (EDCs) in runoff from a potato field irrigated with treated wastewater in Southern California. *Journal of Health Science, 55,* 306–310. https://doi.org/10.1248/jhs.55.306.

Yilmaz, B., Sandal, S., & Carpenter, D. O. (2012). PCB 9 exposure induces en-dothelial cell death while increasing intracellular calcium andROS levels. *Environmental Toxicology, 27*(3), 185–191.

Yoshimura, H., Yoshihara, S. I., Ozawa, N., & Miki, M. (1979). Possible correlation between induction modes of hepatic enzymes by PCBs and their toxicity in rats. *Annals of the New York Academy of Sciences, 320,* 179–192.

Yoshimura, H., Yoshihara, S. I., Koga, N., Nagata, K., Wada, I., Kuroki, J., & Hokama, Y. (1985). Inductive effect on hepatic enzymes and toxicity of congeners of PCBs and PCDFs. *Environmental Health Perspectives, 59,* 113–119.

Yu, G., Sun, D., & Zheng, Y. (2007). Health effects of exposure to natural arsenic in groundwater and coal in China: An overview of occurrence. *Environmental Health Perspectives, 115*(4), 636–642.

Yutani, A., Nakatani, T., Ozaki, A., Yamaguchi, Y., & Yamano, T. (2022). Survey of total mercury content in fishery products. Shokuhin Eiseigaku zasshi. *Journal of the Food Hygienic Society of Japan, 63*(2), 85–91.

Zoeller, R. T., Brown, T. R., Doan, L. L., Gore, A. C., Skakkebaek, N. E., Soto, A. M., Woodruff, T. J., & Vom Saal, F. S. (2012). Endocrine-disrupting chemicals and public health protection: A statement of principles from the endocrine society. *Endocrinology, 153*(9), 4097–4110.

Chapter 4
Interactions, Degradation, and Removal of Endocrine-Disrupting Chemicals

Abstract Endocrine-disrupting chemicals (EDCs) interfere with nuclear hormone receptors (NHRs) and alter hormonal balance, posing a serious risk to human and environmental health. Hydrophobicity and functional groups are two examples of the physicochemical characteristics that determine their persistence, bioaccumulation, and biomagnification. A number of remediation techniques have been investigated to lessen EDCs pollution, including physical, chemical, and biological methods. This study underscores the importance of integrated chemical, biological, and nanotechnological approaches for effective EDCs remediation. Chemically advanced oxidation processes, particularly ozonation and photocatalysis using ZnO and TiO_2-based nanomaterials, have proven effective in degrading persistent EDCs. Additionally, hybrid methods, such as Fenton-based reactions and LED-assisted photo-Fenton techniques, have shown promising results in wastewater treatment. Bacterial strains such as Pseudomonas putida and Bacillus sp. exhibit high EDCs degradation potential, while phytoremediation using species like Lemna minor and Brassica juncea is also used in the removal of EDCs. Despite advancements in these technologies, challenges remain in achieving complete EDCs removal due to their persistence and complex environmental interactions. Future research should focus on optimizing these methods for large-scale applications while ensuring cost-effectiveness and sustainability. Strengthening regulatory frameworks and advancing remediation technologies are crucial to mitigating the long-term impact of EDCs on ecosystems and human health.

Keywords Interactions · Degradation · Removal · Biological methods · Bacteria

4.1 Introduction

Endocrine-disrupting chemicals (EDCs) are biologically active compounds that pose significant threats to environmental integrity. These substances have been associated with both acute and chronic toxicity in aquatic and terrestrial organisms, bioaccumulation in ecosystems, and the deterioration of habitats and biodiversity. Such

M. Kumar et al., *Endocrine Disruptors*,
SpringerBriefs in Molecular Science, https://doi.org/10.1007/978-3-032-15157-5_4

outcomes have raised growing concerns regarding their long-term impact on environmental and ecological health (Jiang et al., 2013; Laurenson et al., 2014). Moreover, EDCs are known to interfere with hormonal regulation by interacting with key biological pathways, leading to a wide range of adverse physiological effects.

The interactions between EDCs and biological systems are intricate. A predominant mechanism involves their affinity for nuclear receptors, which are hormone-dependent transcription factors that orchestrate long-term cellular responses and phenotypic expression (Hwang et al., 2015; Zhang et al., 2022a, 2022b). While EDCs can also influence membrane-bound receptor signaling, these interactions generally result in short-term effects, given the rapid nature of such pathways. Additionally, EDCs may disrupt endocrine regulation by targeting intracellular components downstream of hormone receptor activation rather than directly binding to the hormone receptors themselves. Notably, these molecules often differ structurally from endogenous hormones, yet their disruptive potential remains substantial due to their capacity to alter signaling cascades at various regulatory checkpoints (Combarnous et al., 2019).

From a public health perspective, EDCs have been linked to a multitude of human health issues. These include diminished reproductive and developmental functions, compromised immune responses, increased susceptibility to tumor development, and a heightened risk of neurological disorders (Gao et al., 2020; Kabir et al., 2015). Their toxic nature and complex molecular structures contribute to their environmental persistence and resistance to degradation, posing significant challenges for conventional removal techniques. Several methods have been employed for the removal of EDCs from the environment, including flocculation, precipitation, adsorption, and membrane filtration. However, these methods often demonstrate limited efficiency due to the chemical diversity, low concentrations, and recalcitrant nature of EDCs. As such, more advanced and targeted strategies are being explored to enhance removal efficiency. These include chemically advanced oxidation processes, biodegradation, photocatalytic degradation, and high-efficiency adsorption techniques, which offer promising avenues for mitigating EDC contamination in aquatic environments.

A comprehensive understanding of the interaction mechanisms and environmental behavior of EDCs is, therefore, essential for developing effective remediation strategies. This chapter focuses on the interactions of EDCs with biological systems and provides a comprehensive overview of their degradation and removal, emphasizing chemical, physical, and biological treatment methods aimed at mitigating their environmental and health impacts.

4.2 Interactions of EDCs and Organisms in Diverse Environments

A wide range of environmental pollutants primarily target nuclear hormone receptors (NHRs), which are involved in interactions with these receptors as part of the molecular processes of EDCs. EDCs have a broad range of effects on an organism's endocrine systems (Tabb et al., 2006; Janošek et al., 2006). These include imitating or opposing the effect of natural hormones or altering their production, metabolism, and transportation. Furthermore, these compounds can function through a variety of routes, such as aryl hydrocarbon receptors, membrane receptors, or the enzymatic machinery involved in hormone manufacturing and metabolism. However, most of the reported harmful effects of EDCs are attributed to their interference with hormonal signaling mediated by nuclear hormone receptors (NHRs) (Toppari et al., 2008; Swedenborg et al., 2009).

According to a report by the European Food Safety Authority (EFSA), EDCs also alter genomic expressions in addition to interfering with "endocrine system binding hormonal" receptors (EFSA, 2013; Gore et al., 2015; Lauretta et al., 2019). Eventually, these changes manifest as symptoms such as different types of cancers, among other conditions (Macedo et al., 2023; Rodgers et al., 2018). In humans' exposure to EDCs typically transpires via contact, ingestion, and inhalation pathways, often resulting in their accumulation within bodily tissues as they progress through the food chain. (Balaguer et al., 2017; Lauretta et al., 2019). EDCs have been shown to perform as "total, partial, or inverted agonists or even as antagonists" inside the human body by binding to endocrine nuclear receptors (NRs) (Gore et al., 2015; Lauretta et al., 2019; Mnif et al., 2011; Schug et al., 2011). According to recent studies, these chemicals target and activate a variety of hormone receptors, including androgen and oestrogenic receptors, aryl hydrocarbon receptors, pregnane X receptor, constitutive androstane receptor, estrogen-related receptor, glucocorticoid receptor, thyroid hormone receptor, retinoid X receptor—AR, ER, AhR, PXR, CAR, ERR, GR, TR, and RXR (Monneret et al., 2017; Laurette et al., 2019).

EDCs can inhibit certain receptor isoforms while activating others (Lubrano et al., 2013; Lauretta et al., 2019). Additionally, EDCs interference with the synthesis, transport, metabolism, and elimination of hormones can also selectively lower the levels of endogenous or native hormones (Lauretta et al., 2019; Mnif et al., 2011). EDCs target several groups or axis of organs and glands in the human body, according to numerous studies. The most notable of these are the hypothalamus-pituitary gland-thyroid (HPT), hypothalamus-pituitary gland-gonads (HPG), and hypothalamus-pituitary gland-adrenal (HPA). Some EDCs affect the CNS, while others can increase the number of adipose tissues (by disruption of energy metabolic homeostasis altering adipose tissue) in the body leading to obesity (Gore et al., 2015; Kabir et al., 2015; Lauretta et al., 2019).

According to a number of studies, up to 80% of the EDCs that are now on the market are "potentially tumorigenic" (Macedo et al., 2023; Rocha et al., 2021; Yoon et al., 2014). Therefore, "endocrine neoplasia" which is defined as uncontrolled cell division is one of the most noticeable outward signs of the interactions of EDCs in the human body. Phthalates were shown to be a 100% risk factor for endocrine neoplasia of the prostate, thyroid, and uterine in the examined publications, and thyroid glands were reported to have the largest positive impact size for cancer risk (Macedo and others, 2023). Another way that EDCs are exposed in the food chain and interact with other species is through the oral route, particularly through food. Two primary ways of bioaccumulation in aquatic organisms are bioconcentration during direct partitioning from the abiotic (water) environment during breathing, and through food sources or trophic transfer (Ruhí et al., 2016).

The hydrophobicity of the chemicals affects trophic transmission and biomagnification, as in higher carnivorous creatures and apex predators like humans (Mackay & Fraser, 2000; Kidd et al., 1995; Ruhí et al., 2016). It's due to the reason that EDCs with higher n-octanol water partition coefficient (log Kow), which preferentially have higher fractionation into organic molecules like lipids in the adipose tissues, have a much higher probability of accumulating in the biota (Mackay & Fraser, 2000; Ruhí et al., 2016). According to the food chain exposure routes, EDCs were divided into four groups, which are EDCs introduced through livestock and agricultural practices, including pesticides used in primary food production (Mantovani et al., 2008), animal feed additives (Mantovani et al., 2009), and hormonal treatments in livestock and poultry (Pirro et al., 2015). Another is the bioaccumulative EDCs, such as dioxins and dioxin-like compounds, which accumulate in food sources highly susceptible to environmental contamination, including dairy products (La Rocca & Mantovani, 2006) and large fatty fish (Mantovani et al., 2016). Third, food contact EDCs like BPA in packages and semicarbazide, leached from heated glass-sealed food, present immediate consumption hazards (Mantovani, 2016). The contaminants also impact pets, which have shared close exposure with humans (Pocar et al., 2023). Lastly, EDCs that occur naturally in the food web are such chemicals as phytoestrogens, which interfere with the reproductive processes of cattle and sheep (Adams, 1995) and affect human beings by way of alcohol produced from phytoestrogen-containing grains (Rosenblum et al., 1993).

These compounds hydrophobicity/hydrophilicity, which is influenced by internal and external factors, including pH, temperature, salinity, cation exchange capacity, clay content, organic matter content, molecular chain length, polarity, and functional groups, determine how these compounds interact in terrestrial and aquatic environments (Xie et al., 2013; Fonseca et al., 2024) (Table 4.1). A study on a group of pharmaceuticals known as selective serotonin reuptake inhibitors (SSRIs) indicated that the -NH2 functional group in fluvoxamine contributed to comparatively lower log Kow (greater solubility) compared to other SSRI class members such as sertraline which contained hydrophobic functional groups such as $-Cl$ and $-CH_3$ (leading to higher fractionation to solid phases such as sediments and soils) (Kwon & Armbrust, 2008). It was also discovered that sorption of hydrophobic and neutral SSRIs on sediments and soil might be related to the organic carbon content of the soil/sediment and the biomass and showed a positive relationship with log Kow. Ionic chemicals, however, showed different mechanisms such as cation exchange, cation bridging at the clay surfaces, surface complexation, and hydrogen bonding during the sorption process (Kwon & Armbrust, 2008).

A study by Zhang et al. (2014) shows positive correlations (correlation coefficients: 0.46–0.57) between particulate EDCs and particulate organic carbon in the North Tai Lake Basin, Eastern China. Numerous studies have examined how chain length affects phytoaccumulation and found that longer-chained, hydrophobic PFASs preferentially partition into solid phases like soils and sediments, while shorter-chained, polar PFASs are more easily absorbed and accumulated through passive transport. This is explained by the energy-intensive nature of the active transport of these compounds in plants through interactions between proteins and lipids (Mayakaduwage et al., 2022). Furthermore, translocation factors (TFs) are lower for longer-chained hydrophobic PFASs, which are identified by greater octanol–water distribution coefficients (D_ow), membrane-water distribution coefficients (D_mw), and organic carbon–water partition coefficients (K_oc). However, because of their increased propensity to fractionate into the neutral, phospholipid, and non-lipid organic matter of aquatic macrophytes, they exhibit larger whole-plant bioconcentration factors (BCFs) despite having restricted mobility within plant tissues.

Table 4.1 The properties (chemical, environmental, retention, transport, accumulation, and transformation) of different EDCs, their type of mechanism, environmental fate, and effects

Process	Key drivers	Chemicals	Outcome/ environmental effect	References
Solubility & partitioning	Ionic strength, polarity	Ionic EDCs, non-polar PFAS	Enhanced mobility or sorption to organic matter	Assmuth & Louekari (2001)
Volatility & stability	Vapor pressure, halogenation	Industrial PFAS, halogenated pesticides	Long-range atmospheric transport	Kwon & Armbrust (2008)
Environmental conditions	pH, salinity, temperature	Ionic plasticizers	Affect ionization, hydrophobicity, and adsorption	Borrirukwisitsak et al. (2012)
Transport Pathways	Sorption, advection, diffusion	PhACs, EDCs	Movement through water, air, and soil	Kinney et al. (2006)
Transformation	Hydroxylation, methylation, photolysis	PCBs, metals	Alters persistence and toxicity	Caliman & Gavrilescu (2009)
Bioaccumulation	Lipophilicity (log Kow)	Persistent EDCs	Accumulation in biota and sediments	Pi et al. (2017)

4.3   Removal and Degradation of Endocrine-Disrupting Chemicals

EDCs pose a hidden threat to both the environment and people. This situation can be readily managed if EDCs can be eliminated from sewage at sewage treatment facilities before ultimate discharge into the environment. Currently, there are three main techniques for treating EDCs: physical methods, chemical methods, and biological treatments.

4.3.1   Physical Methods

Activated carbon (AC) is a widely recognized method for eliminating a variety of organic pollutants. In packed bed filters, AC is most frequently used as a powdered feed (powder-activated carbon, or PAC) or in granular form (granular activated carbon, or GAC). Using a GAC fixed bed, removal performances for three EDCs (bisphenol A, amylphenol, and nonylphenol) were investigated on a laboratory scale. Data indicated that EDCs with high Kow (octanol–water partition coefficient) values may be removed successfully. The authors also discovered that the varieties of activated carbon and their service life had varying effects on the removal of EDCs. The coal-based carbon was shown to be more successful because of its bigger pore volume, which was demonstrated to be more related to AC's absorbability than its specific area (Choi et al., 2005). The adsorbing ability of AC has also been enhanced by modification; Vidal et al. (2016) found that adding sulfur to the carbon structure of AC increased the dynamic adsorption of trimethoprim from 195 mg/g to 240 mg/g. In comparison to other adsorbents such as carbon nanotubes (103.81 mg/g), graphene oxide (77.86 mg/g), and biochar (9.19 mg/g), PAC has demonstrated the highest removal efficiency (132.73 mg/g) for oestrogen. It has been established that AC treatment is extremely effective in the removal of nearly all target chemicals (Jiang et al., 2013). Highly hydrophobic EDCs, such as pharmaceuticals and personal care products (PPCPs), have shown up to 75% removal efficiency when activated carbon (AC) is incorporated into secondary filtration. Notably, acebutolol, diazepam (DZP), and diltiazem were eliminated during treatment (Fu et al., 2019). Another study by Rao et al. (2021) reported 90–98% removal of PPCPs at lower concentrations, with extended contact time further enhancing removal efficiency (Table 4.2).

In a study sedimentation process was used to get rid of EDCs such as E1, ibuprofen, and sulfamethoxazole (Datel & Hrabankova, 2020). Features of the EDCs, such as the acid solubility (pKa), which establishes the solubility of the chemical based on surface charge, and the octanol–water partition coefficient (log K_{ow}), are crucial for the success of sedimentation (Zamri et al., 2021). Therefore, compounds with larger log K_{ow} values are easier to separate, which implies increased fractionation into the sediment fraction. Consequently, it has been suggested that sedimentation is often ineffective as a technique for removing EDCs.

Table 4.2 The physical, chemical, and biological treatment types of EDCs. Various degradation technologies and approaches used to treat EDCs in different media, with their removal efficiencies, are mentioned

Treatment type	Physical			
Technology used	EDCs treated	Media	Removal efficiency (%)	References
AC	TCS, naproxen, ibuprofen, and ketoprofen	Wastewater	• TCS:84, • Naproxen:91, • Ibuprofen:95, • Ketoprofen:93	Noutsopoulos et al. (2014)
AC	PPCPs	Wastewater	• Up to 75% (acebutolol, diazepam, and diltiazem were completely removed during the treatment process)	Fu et al. (2019)
β-CD@MRHC (Modified cellulose)	Pb (II) and BPA	Laboratory experiment (aqueous solution)	• Superior adsorption performance	Liu et al. (2022)
CNTs	Wide range of EDCs	Laboratory experiment (aqueous solution)	• A wide range of EDCs have been removed but real-world applications are lacking	Kuwadkar et al. (2019)
GO	E2	Laboratory experiment (aqueous solution)	• Maximum adsorption capacity: up to 149.4 mg/g	Jiang et al. (2013)
Modified cellulose	BPA	Laboratory experiment (but with many real-life applications)	• At ultrapure water:77 • Synthetic water: 64	Tursi et al. (2018)
MBR-integrated RO system	PPCPs	Wastewater	• >95	Wang et al. (2018)

(continued)

Table 4.2 (continued)

Treatment type	Physical			
Technology used	EDCs treated	Media	Removal efficiency (%)	References
Modified NF (NF membrane intercalated with hydrophilic MoS$_2$ nanosheets, and MOF thin-film Nanocomposite NF membrane)	Parabens and BPA	Laboratory experiment (aqueous solution)	• The removal efficiency for all types of EDCs was improved by the two different NF membrane modifications	Dai et al., (2021a, b)
NF	Tetracycline and sulphonamides	Laboratory experiment (dead-end filtration cell)	• Chlortetracycline adsorbed: 80 • Doxycycline: 50 • Hormones: lower than tetracyclines	Koyuncu et al. (2008)
Sedimentation	DCF and E3	Wastewater	• ≤28	Behera et al. (2011)
Sedimentation	E1, ibuprofen, and sulfamethoxazole	Wastewater	• Low efficiency	Gao et al. (2012), Datel & Hrabankova (2020)
UF	Amoxicillin, naproxen, metoprolol, and phenacetin	Laboratory experiment (aqueous solution)	•. Works on hydrophobic EDCs	Ojajuni et al. (2015), Patel et al. (2019)

(continued)

Table 4.2 (continued)

Treatment type	Physical			
Technology used	EDCs treated	Media	Removal efficiency (%)	References
UF in combination with other techs	Estrogen analogs	Laboratory experiment (aqueous solution)	• Combination with GAC, magnetic ion exchange resin, and ozonation can result in more effective removal	Huang et al. (2019)
US	Acetaminophen and naproxen (PPCPs)	Laboratory experiment (aqueous solution)	• Naproxen: >99 • Acetaminophen: 86.1	Im et al. (2014)
US	TCS, erythromycin, and iopromide	Laboratory experiment (aqueous solution)	• TCS: > 95 for TCS • Erythromycin and iopromide: Partial	Naddeo et al. (2013)
Chemical				
Chlorination, ozonation, permanganate, and chloramine	EDCs and pharmaceuticals (TCS, ibuprofen, clofibric acid, E1, E2, E3, iopromide and EE2)	Laboratory experiment (aqueous solution)	• Chlorination, ozonation, and permanganate all were successful in the removal of E3, TCS, and • E1, EE2 (not chlorination) but not ibuprofen, iopromide, and clofibric acid • Chloramine: not very effective	Wu et al. (2012)

(continued)

Table 4.2 (continued)

Treatment type	Physical			
Technology used	EDCs treated	Media	Removal efficiency (%)	References
Fenton like processes	E2 and EE2	Laboratory experiment (aqueous solution)	• E2: 95 & EE2: 98 at ferric concentration of 1×10^{-3} M (58.6 mg/L)	Ifelebuegu & Ezenwa (2011)
LED-irradiated photo-Fenton method	4-octylphenol, 4-n-NP, BPA, E1, E2, EE2, and E3	Wastewater	• Estrogenic activity: 62 • Mineralization: 59	Silva et al., (2021)
Ozonation	BPA	Laboratory experiment (aqueous solution)	• Dose-dependent behavior-Degradation: 70, 82, 90 at 1 mg/L, 1.5 mg/L, 2 mg/L, respectively, at exposure of 30 min	Xu et al. (2006)
Ozonation and chlorination	55 EDCs/PPCPs	Wastewater	• Ozonation of carbamazepine and losartan: > 99 • Chlorination: 99 • Venlafaxine and its demethylated metabolite, carbamazepine epoxide, phenytoin, and hydrochlorothiazide: partially removed	Huerta-Fontela et al. (2011)

(continued)

Table 4.2 (continued)

Treatment type	Physical			
Technology used	EDCs treated	Media	Removal efficiency (%)	References
Photocatalytic oxidation	BPA, BPB, diamyl phthalate, BBP, methyl p-hydroxybenzoate, and ethyl 4-hydroxybenzoate	Experimental (solar pilot plant)	• Dissolved organic carbon: 83 • Degradation trend: BPs >parabens >phthalates • Concentration reduced to 0% for BPB ($<$ LOQs)	Vela et al. (2018)
Photooxidation in the presence of $Ce_xZn_{1-x}O$ nano-photocatalyst	BPA	Laboratory experiment (aqueous solution)	• Effective than pure ZnO catalysts. Factors affecting the degradation: BPA concentration, catalyst load, exposure time, and pH value	Kamaraj et al., (2014)
Photooxidation (visible light) with rGO-Ag-based composite	Phenol, BPA, and atrazine	Laboratory experiment (aqueous solution)	• Degradation was efficient and better than individually using rGO or silver nanoparticles	Bhunia & Jana (2014)
Photooxidation ($CuInSe_2$/ $NiFe_2O_4$ nanocomposite)	BPA and resorcinol	Laboratory experiment (aqueous solution)	• Increase in $NiFe_2O_4$ content (35%): increase in degradation rate • $NiFe_2O_4$ or $CuInSe_2$ individually: least effective	Jiang et al. (2021)
Biological				

(continued)

Table 4.2 (continued)

Treatment type	Physical			
Technology used	EDCs treated	Media	Removal efficiency (%)	References
Immobilized-laccase-based catalyzed system	E1, E2 and EE2	Laboratory experiments (packed-bed reactor system)	• Methyltrimethoxysilane and tetraethoxysilane combination sol–gel matrix for immobilization was effective in degrading >85% of E1, E2, and EE2	Barrios-Estrada et al., (2018)
Laccase-catalyzed system	NP, octylphenol, BPA and EE2	Laboratory experiments	• Laccase obtained from the fungi *Trametes versicolor*, *Trametes villosa*, *Coriolopsis polyzona* have been successful in treating BPA; • Laccase from *Trametes sp.* (Laccase Daiwa) was found to be effective against the treatment of phenolic EDCs like NP, octylphenol, BPA, and EE2	Tanaka et al., (2001), Kim & Nicell (2006), Barrios-Estrada et al., (2018)

(continued)

Table 4.2 (continued)

Treatment type	Physical			
Technology used	EDCs treated	Media	Removal efficiency (%)	References
Laccase mediator-catalyzed system	Various EDCs	Laboratory experiments	• Mediators like Triton X-100 enhance the removal rates of aqueous phenols • Mediator addition: improved the degradation of BPA and DCF from a maximum of >85% and >60% respectively, (only laccase) to >95% and >80, respectively	Zhang et al., (2012), Barrios-Estrada et al., (2018)
Microorganisms: *Sphingomonadales* and *Rhodococcus* used in two-phase partitioning bioreactor	E1, E2, E3, EE2, NP, and BPA	Wastewater	• Effective degradation, except in the case of EE2	Villemur et al., (2013)
Microorganisms: YC-AE1 (*Pseudomonas putida*)	BPA	Laboratory culture	• Degradation: BPA (99%) at 500 mg/L within 72 h • Degrade related pollutants like BPB, BPF, BPS, DBP, DEHP, and DEP	Eltoukhy et al., (2020)

(continued)

Table 4.2 (continued)

Treatment type	Technology used	EDCs treated	Media	Removal efficiency (%)	References
Physical	Microorganisms: *Bacillus sp.* MY156	DEHP	Batch experiments	• Complete rate of degradation of DEHP at 5 days of growing the strain at a pH 8 at 30 °C	Xie et al. (2023)
	Nature based solutions: phytoremediation	PFASs	Removal assessed in naturally growing plants	• Effective removal of PFAs observed for *Betula pendula, Picea abies, Prunus padus, Sorbus aucuparia, Aegopodium podagraria, Phegopteris connectilis,* and *Fragaria vesca*	Gobelius et al., (2017)
	Nature-based solutions: phytoremediation via wetland plants	PFOA	Macrophytes collected randomly from wetlands	• Removal efficiency for PFOA in decreasing order *Eichhornia crassipes > Polygonum salicifolium > Cyperus congestus > Persicaria amphibian > Ficus carica > Artemisia schmidtiana > Xanthium strumarium > Phragmites australis > Ruppia maritime > Schoenoplectus corymbosus*	Mudumbi et al., (2014)

According to Mallakpour & Soltanian (2016), there are two types of carbon nanotubes (CNTs): single-walled (SWCNTs), which have a single rolled graphite sheet layer, and multiwalled (MWCNTs), which have two or more rolled graphite sheet layers. The adsorption location, purity, and surface functional groups of nanometer-thick layered carbon, as well as the pH, ionic strength, starting solute concentration, and temperature, all have an impact on CNT adsorption in contrast to AC (Arora & Attri, 2020). CNTs have been used to remove a wide range of EDCs, including tetracycline (175 mg/g), diuron (40.37 mg/g), BPA (92 mg/g), and E2 (27.2 mg/g) (Kuwadkar et al., 2019; Martín-Lara et al., 2020).

Cellulose has shown remarkable efficacy in eliminating EDCs from aqueous solutions when compared to synthetic adsorbents (Zamri et al., 2021). However, because of the limited repeatability of the adsorption process, native cellulose is often less successful in removing aquatic EDCs and pollutants of rising concern. However, Tursi et al. (2018) found that bio-based Spanish broom (SB) surface-modified cellulose fibers removed 77% of BPA in ultrapure water and 64% in synthetic water, demonstrating the improved adsorption ability of modified cellulose. A single treatment technique might not be enough since water bodies frequently include a variety of pollutants. Recently, in a study by Liu et al. (2022) recently showed how well β-cyclodextrin-modified magnetic rice husk-derived cellulose (β-CD@MRHC) removed BPA and Pb (II), a trace metal ion, at the same time with high recovery rates. Notably, because Pb (II) and BPA have different capture mechanisms-ion exchange, complexation, and electrostatic attraction for Pb (II) and hydrogen bonding and host–guest inclusion for BPA-β-CD@MRHC was able to successfully separate their competing contacts (Liu et al., 2022).

Membrane technologies (MT) have also advanced significantly and offer enormous promise for removing numerous pollutants, including EDCs, from a variety of media, such as wastewater. In MT, reverse osmosis (RO), nanofiltration (NF), and ultrafiltration (UF) are the various pore-size-based techniques. The removal of EDCs from WWTPs has seen widespread real-world applications of RO technology. In particular, RO has successfully eliminated BPAs, PPCPs, and pesticides (Zamri et al., 2021; Ahmad et al., 2022). In a WWTP, the removal efficiency of EDCs resulting from an integrated system combining membrane bioreactor (MBR) and RO was higher than 95%; in fact, the removal efficiency of the MBR system itself was as high as 95%. This indicates that integrated systems and modified RO systems are even more effective than standard RO systems (Wang et al., 2018).

Another MT that has been shown to eliminate EDCs such as hormones and PPCPs from water is NF (Koyuncu et al., 2008). The pores are between 0.2 and 0.4 nm in size, and pollutants are eliminated via size exclusion and adsorption on the membrane surface. Monovalent ions and water are permitted to flow through but divalent and multivalent ions are kept, indicating charge selectivity in permeability even for dissolved components (Dharupaneedi et al., 2019). Comparing modified nanomembranes to a control conventional polyamide (PA) nanofiltration (NF) membrane, it was discovered that the latter modified by MoS_2 nanosheets (NS) increased the removal of hydrophobic EDCs from wastewater (Dai et al., 2021a). Therefore, this alteration led to increased EDCs rejection rates because the

MoS$_2$-NS-induced nanochannels had a selective impact and the hydrophobic contact between the membrane surface and the EDCs was suppressed. The rejection rates of hydrophobic EDCs may also be increased by manipulating water transport nanochannels by increasing the density of the polyamide layer in the metal–organic framework (MOF)-integrated thin-film nanocomposite (TFN) membranes (Dai et al., 2021b).

In addition to its benefits, US, is another method of degrading contaminants, which can act on complex contaminants in WWTPs as well as low concentrations (ng/L to µg/L levels) of contaminants (Zamri et al., 2021). The US works by causing thermal and chemical reactions in both gaseous and aqueous solutions (Zamri et al., 2021). In a study by Im et al. (2014), the degradation of acetaminophen and naproxen (PPCPs) at a frequency of 1000 kHz was >99% for naproxen and 86.1% for acetaminophen (Zamri et al., 2021). In a study by Park et al. (2011) Sono degradation of common EDCs (BPA, E2, and EE2) was carried out at various frequencies. The findings showed that the sonication frequency had a significant impact on the breakdown of EDCs since a higher frequency generally resulted in more free radicals in the system. The degradation by the US is also influenced by the contaminant's chemical structure (Naddeo et al., 2013; Zamri et al., 2021). TCS was shown to be up to 95% reduced after 180 min of ultrasonication treatment, however, other pollutants such as erythromycin and iopromide (IPM) were only slightly degraded (Naddeo et al., 2013). EDCs retention by the membrane processes is mainly due to size exclusion, charge repulsion, and adsorption. In comparing membrane types over most cases, EDCs rejection rate by reverse osmosis is the highest, followed by nanomembrane types, then ultra-membranes, with the rejection of micro-membranes as the lowest.

4.3.2   Chemical Methods

Chemical advanced oxidation (CAO) is the process of removing EDCs using various chemical oxidants. The primary processes of CAO include the mineralization of wastewater contaminants to CO_2 or the transferring of pollutants to other metabolite products via oxidation–reduction reactions involving strong oxidizers. It was discovered that the oxidative processes of ozonation and chlorination were far more effective in breaking down or eliminating EDCs. In a study, four distinct oxidative techniques chlorination, ozonation, permanganate, and chloramine were used to treat pharmaceuticals and EDCs (TCS, ibuprofen, clofibric acid, E1, E2, E3, IPM, and EE2). It was observed that E3, TCS, and E1 were shown to be very effectively removed by chlorination, however, ibuprofen, IPM, and clofibric acid were not. Permanganate was inefficient at eliminating ibuprofen, clofibric acid, and IPM, while it was very successful at eliminating TCS, E1, EE2, and E3. The substances ibuprofen, clofibric acid, and IPM were the only EDCs that could be removed by ozonation, but the EDCs and pharmaceuticals could not be efficiently removed by chloramine at the standard disinfection dosage of about 3 mg/L (Wu et al., 2012). It was observed that initial BPA concentration, dose, water background, and ozone exposure duration are some of the variables that affect ozonation efficiency for BPA degradation (Xu et al., 2006). While higher doses of ozone effectively degraded BPA and removed byproducts, lower doses were ineffective.

Advanced oxidation technologies are of different types among which photocatalytic technology is regarded as the most promising of these water treatment techniques because of its high rate of degradation and capacity for mineralization (Guo et al., 2017). Using sunlight as the excitation energy, photocatalytic technology may mineralize and break down stubborn pollutants (such as organic and inorganic materials) into CO_2, H_2O, and a trace amount of simple inorganic molecules. Six distinct EDCs (BPA, BPB, diamyl phthalate, BBP, methyl p-hydroxybenzoate, and ethyl 4-hydroxybenzoate) were broken down in a pilot plant using photooxidation. The degradation rates were found to be significantly enhanced by the addition of ZnO as a photocatalyst and $Na_2S_2O_8$ as an electron acceptor. The sequence of degradation was found to be bisphenols > parabens > phthalates (with 0% concentration or <LOQ in the case of BPB). In the bacterium Vibrio fisheri, this resulted in a reduction in toxicity to tolerable levels of 11% inhibition and a drop in dissolved organic carbon to 83% of the initial value. It has been demonstrated that the TiO_2-based nanofiber powder photocatalyst NnF Ceram TiO_2 easily breaks down progesterone and all forms of oestradiol (Zatloukalová et al., 2017).

Advanced oxidation processes (AOP) are oxidative techniques that have been improved by using nanomaterials as catalysts to increase the efficiency of visible light capturing, as well as the generation and more efficient utilization of electrons and holes, and the prevention of their recombination (González-González et al., 2022). Using techniques like doping transition metal ions, depositing precious metals, and coupling with other semiconductors like CeO_2–ZnO, ZnO–WO_3, and TiO_2-ZnO, to stop the recombination of charge carriers in semiconductors and boost the photocatalytic efficiency of ZnO catalysts (Pant et al., 2012). Photooxidation under sunlight when used in the presence of $Ce_xZn_{1-x}O$ nano-photocatalyst was found to have higher effectiveness in the removal of BPA compared to the rate observed in the presence of pure ZnO (Kamaraj et al., 2014). $CuInSe_2/NiFe_2O_4$ nanocomposites (solid state preparation by simple MOF-assisted synthesis at moderate temperatures) were found to be highly effective when utilized during the photocatalytic oxidation of the EDCs BPA and resorcinol (Jiang et al., 2021).

$Ce_xZn_{1-x}O$ nanocomposites were synthesized by the coprecipitation method. The prepared photocatalyst exhibited higher photocatalytic efficiency than pure ZnO during BPA degradation under sunlight irradiation and achieved almost complete BPA mineralization, indicating $Ce_xZn_{1-x}O$ assisted photocatalytic degradation is an economical, environmentally friendly, and effective approach for removing BPA in an aqueous system (Gao et al., 2020). ZnO nanorods (ZNRs) were coated with nanoscale Bi_2O_3 particles using a combination of chemical precipitation and hydrothermal technologies. The findings demonstrated that the tiny Bi_2O_3 nanoparticles were dispersed uniformly over the ZNRs' surface. Furthermore, the excellent charge separation efficiency and •OH production ability of the Bi_2O_3–ZNR nanocomposites were observed. Therefore, with the elimination of two EDCs, phenol, and methylparaben, the Bi_2O_3–ZNR nanocomposites demonstrated more photocatalytic activity than a pure ZNR catalyst. Furthermore, the one-dimensional nanostructure features of the

Bi_2O_3–ZNR nanocomposites made them simple to recycle and reuse (Lam et al., 2013).

When used in the photocatalytic oxidation of the EDCs BPA and resorcinol, $CuInSe_2/NiFe_2O_4$ nanocomposites (solid state fabrication via simple MOF-assisted synthesis at low temperatures) were found to be highly effective (Jiang et al., 2021) (Table 4.2). The main advantage was that the $CuInSe_2/NiFe_2O_4$ nanocomposites could be recycled after being magnetically extracted from treated wastewater by an external magnetic field. In laboratory-scale tests, other oxidation techniques, such as Fenton-type reactions, have also been used to break down EDCs like E2 and EE2, with removal efficiencies of 95% and 98%, respectively (Ifelebuegu & Ezenwa, 2011) (Table 4.2). Silver orthophosphate (Ag_3PO_4) due to its ability to absorb visible light can effectively degrade organic pollutants. $LaCoO_3$, a perovskite-type transition metal oxide with magnetic properties, is also gaining attention due to its high catalytic activity, low costs, and no environmental pollution. A study showed that $Ag_3PO_4/LaCoO_3$ composites have potential environmental applications, with approximately 77.27% of BPA mineralized and degraded under 40 min of irradiation. These photo-catalysts are stable and reusable (Guo et al., 2017). Therefore, a more advanced LED irradiated photo-Fenton method used for the treatment of 4-octylphenol, 4-n-NP, BPA, E1, E2, EE2, and E3 resulted in 62% removal of estrogenic activity and 59% mineralization in real wastewater treatment plant effluent (WWTPE) (Silva et al., 2021) (Table 4.2). In this study, additionally, the treated WWTPE was found to be non-toxic to Aliivibrio fischeri, and also >80% of the EDCs were removed.

4.3.3   Biological Methods

Biological methods are designed to assess the concentration of specific EDCs, assuming that the target compound has already been identified as an endocrine disruptor and its chemical properties are well understood. The degradation of EDCs using biological agents like microbes has been explored with the progress in microbiology and biotechnology. Villemur et al. (2013) used microbial degradation. In a study to remove EDCs from WWTPs, a two-phase partitioning bioreactor (TPPB) was developed. It was observed that only EE2 remained adequately undergraded in the enrichment cultures out of E1, E2, E3, EE2, NP, and BPA. (Villemur et al., 2013). Another study by Eltoukhy et al., (2020) found to have high degradation potential for BPA over a wide range of concentrations (0.5–1000 mg/L), with complete degradation at 500 mg/L within 72 h using YC-AE1 (Pseudomonas putida), a Gram-negative strain isolated from soil. In the same study it was also found after substrate analysis that YC-AE1 could also degrade a number of other related organic pollutants like BPB, BPF, BPS, DBP, DEHP, and DEP.

A recent study by Xie et al. (2023) investigated the biodegradation of di(2-ethylhexyl) phthalate (DEHP) using a bacterial strain, Bacillus sp. MY156, isolated from wastewater treatment plant (WWTP) sludge (Table 4.2). The study reported a nearly complete degradation of DEHP at an initial concentration of 500 mg/L within

5 days under optimized conditions of pH 8 and 30 °C. The presence of trace element ions, including Fe^{3+}, Zn^{2+}, and Mn^{2+}, was found to enhance the degradation process, whereas Cu^{2+} exhibited an inhibitory effect. Identified degradation intermediates included mono-ethylhexyl phthalate (MEHP), diethyl phthalate (DEP), dimethyl phthalate (DMP), phthalic acid (PA), and protocatechuate. The findings suggest that Bacillus sp. MY156 possesses significant potential for the biodegradation of phthalic acid esters (PAEs) (Xie et al., 2023).

Laccases are oxidative enzymes naturally found in various species, including bacteria, insects, vascular plants, and fungi. Their use has been the foundation of the biological oxidative process (Barrios-Estrada et al., 2018; Gasser et al., 2014; Jones & Solomon, 2015; Madhavi & Lele, 2009). For the degradation of EDCs, laccases have been used in three distinct ways: (1) a system catalyzed by laccase mediators, (2) degradation using a laccase-catalyzed system, and (3) degradation using immobilized laccase-based catalyzed system (Barrios-Estrada et al., 2018). A wide range of EDCs has been treated by employing a laccase-catalyzed system, which uses laccase directly as a catalyst (Barrios-Estrada et al., 2018).

The ability of laccase derived from white rot fungal strains such as Trametes versicolour, Trametes villosa, Coriolopsis polyzona, to remove BPA has been tested; in particular, the laccase derived from Trametes sp. (Laccase Daiwa) was found to be effective in breaking down phenolic EDCs such as NP, octylphenol, BPA, and EE2 (Barrios-Estrada et al., 2018; Tanaka et al., 2001). However, the non-reusability of free laccase-catalyzed systems and their inactivation are drawbacks of using laccase directly as a catalyst. These issues have been tackled with the addition of chemicals that mediate the laccase-catalyzed systems, some EDCs with higher redox potential than that expressed by laccase need these mediators. Triton X-100 A mediator was able to improve the elimination of aqueous phenols catalyzed by laccase (Zhang et al., 2012). Therefore, the inclusion of a natural redox-mediator molecule (syringaldehyde, 5 µM) increased the efficiency of a recently developed laccase-based enzymatic membrane reactor (EMR) from a maximum of >85% and >60% of BPA and DCF removal to >95% and >80%, respectively.

Phytoremediation is also a potential method for removing EDCs from the environment using various plant species. Different plant species, including Ceratophyllum demersum, Myriophyllum spicatum, Cannabis sativa, Chrysopogon zizanioides, Brassica juncea, Lolium perenne, Landoltia punctata, and Lemna minor, were evaluated for their ability to remove contaminants such as bisphenol A (BPA), metalaxyl-M, tetracycline (TTC), ciprofloxacin (CIP), mercury (Hg), atrazine, methylparaben (MeP), and propylparaben (PrP). The removal efficiency varied depending on the plant species and contaminant, with BPA being removed up to 100% by Ceratophyllum demersum and Myriophyllum spicatum (Zhao et al., 2021), while Cannabis sativa showed high removal efficiency for both BPA (86%–95%) and metalaxyl-M (67%–94%) (Loffredo et al., 2021). Another study by Panja et al., (2020) demonstrated the effective removal of CIP (60%–94%) and TTC (89%–100%) using Chrysopogon zizanioides.

Several bacterial strains, including Ochrobactrum sp., Pseudomonas sp., Acinetobacter sp., Lysinibacillus sphaericus, Bacillus, Desulfobacter, and Pseudomonas putida, were also evaluated for their ability to degrade various EDCs such as E2, BPA, methylparaben (MPB), butylparaben (BPB), ciprofloxacin (CIP), and 17α-ethinylestradiol. The removal efficiencies varied, with Ochrobactrum sp. strain FJ1 achieving 98% removal of E2 in 10 days (Zhang et al., 2022a, 2022b) and Pseudomonas sp. (GBPI_508) removing 97% of BPA (Thathola et al., 2022). While Acinetobacter sp. and Pseudomonas sp. exhibited lower BPA degradation (20–36%) (Noszczyńska et al., 2021). Enzymes such as dehydrogenase, esterase, peroxidase, cytochrome P450, and manganese oxide also played key roles in degradation. These findings highlight the potential of bacterial strains in EDCs bioremediation, with variations in efficiency based on strain, contaminant type, and enzymatic pathways.

Constructed wetlands (CWs) or hyperaccumulating plantations in the soil are also examples of nature-based solutions that are utilized to repair and remediate pollutants like PFASs from soil and water (Mayakaduwage et al., 2022). According to a Swedish study, PFASs can be removed from the soil by bioaccumulation from plants such as silver birch (Betula pendula), Norway spruce (Picea abies), bird cherry (Prunus padus), mountain ash (Sorbus aucuparia), ground elder (Aegopodium podagraria), long beech fern (Phegopteris connectilis), and wild strawberry (Fragaria vesca) (Gobelius et al., 2017). In this study, the bioconcentration factors (BCFs; plant/soil ratios) were found to be the highest in the foliage. Agricultural plants like tomato (Solanum lycopersicum), cabbage, and zucchini (Cucurbita pepo) grown hydroponically were utilized to remove a number of different PFASs (n = 14), it was found that shorter chained PFASs (<C11) were translocated more easily than longer chained members (Felizeter et al., 2014). In common wetland macrophyte species, the accumulation of PFOA (in a concentration range of 11.7–38 ng/g) was in the order of Eichhornia crassipes > Polygonum salicifolium > Cyperus congestus > Persicaria amphibian > Ficus carica > Artemisia schmidtiana > Xanthium strumarium > Phragmites australis > Ruppia maritime > Schoenoplectus corymbosus (Mudumbi et al., 2014).

In plants four different mechanisms are observed during phytoremediation of EDCs like PFASs; phytostabilization involves the immobilization or stabilization of both inorganic as well as organic pollutants in the soil (Herath & Vithanage, 2015). Stabilization is achieved by adsorption on roots, adsorption, and accumulation by roots and by precipitation within the rhizosphere (Awa & Hadibarata, 2020). Phytoextraction/phytoaccumulation is a mechanism involving the uptake of pollutants by the roots from both soil and water, the translocation of the compounds to above-ground biomass, and the accumulation in harvestable tissues (Gunarathne et al., 2020). The transformation of the contaminants into volatile form and release into the atmosphere after transportation through the xylem is called phytovolatilization (Ali et al., 2013). Finally, phytodegradation involves the breakdown of the contaminants into less toxic forms after accumulation within the plants (Huang et al., 2021). In a study conducted by Gong et al (2018), after phytoremediation (by Boehmeria nivea (L.)) of trace elements in contaminated sediments, the plant biomass was further pyrolyzed at 300–700 °C, which stabilized the contaminants, and the resultant biochar was used for the adsorption of an organic pollutant (dye). This integration of phytoremediation and adsorption post-pyrolysis also applies the concept of circular economy by reusing plant biomass and has the potential for wide-scale application.

The number of techniques discussed was experimental, as the real-life applications for physical, chemical, and biological approaches lagged significantly compared to laboratory applications. Although the number of studies reviewed for each of the categories was not similar, it could be observed that physical methods were the most widely used and had the highest success rates (33.33%, 29%, and 4.76% for practical applications in physical, chemical, and biological methods, respectively). A common observation was that the removal of estrogenic chemicals was inefficient in both physical as well as chemical methods.

4.4 Conclusion

EDCs are persistent environmental contaminants that interfere with endocrine regulation and have severe consequences for human health and the environment. Bioaccumulation and biotransfer demand the use of efficient remediation technologies. Of all the remediation technologies, chemical advanced oxidation processes like ozonation and photocatalytic degradation are highly efficient in the degradation of recalcitrant contaminants. The use of nanoparticles improves these processes even further by reducing secondary pollutants and increasing degradation efficiency. With the potential for widespread use, biological techniques such as phytoremediation and microbial degradation provide ecologically benign and sustainable substitutes. A comprehensive remediation approach that incorporates chemical, biological, and nanotechnological treatments is necessary due to the intricacy of EDCs contamination. Furthermore, developments in adsorption and membrane filtering technologies have the potential to increase removal efficiency. To maximize these methods for real-world use while maintaining cost-effectiveness, more study is necessary. Strict regulatory measures and the advancement of environmentally friendly remediation technology must be given top priority to lessen the long-term effects of EDCs and safeguard environmental integrity and public health.

References

Adams, N. R. (1995). Detection of the effects of phytoestrogens on sheep and cattle. *Journal of Animal Science, 73*(5), 1509–1515.

Ahmad, N. A., Goh, P. S., Azman, N., Ismail, A. F., Hasbullah, H., Hashim, N., Kerisnan Krishnan, N. D., Yahaya, N. K. E. M., Mohamed, A., Mohamed Yusoff, M. A., Karim, J., & Abdullah, N. S. (2022). Enhanced removal of endocrine-disrupting compounds from wastewater using reverse osmosis membrane with titania nanotube-constructed nanochannels. *Membranes, 12.* https://doi.org/10.3390/membranes12100958.

Ali, H., Khan, E., & Sajad, M. A. (2013). Phytoremediation of heavy metals—Concepts and applications. *Chemosphere, 91*(7), 869–881.

Arora, B., & Attri, P. (2020). Carbon nanotubes (CNTs): A potential nanomaterial for water purification. *Journal of Composite Science.* https://doi.org/10.3390/jcs4030135.

Assmuth, T., & Louekari, K. (2001). Research for management of environmental risks from endocrine disrupters. Helsinki: Finnish Environment Institute.

Awa, S. H., & Hadibarata, T. (2020). Removal of heavy metals in contaminated soil by phytoremediation mechanism: A review. *Water, Air, & Soil Pollution, 231*(2), 47.

Balaguer, P., Delfosse, V., Grimaldi, M., & Bourguet, W. (2017). Structural and functional evidences for the interactions between nuclear hormone receptors and endocrine disruptors at low doses. *Comptes Rendus. Biologies, 340*(9–10), 414–420.

Barrios-Estrada, C., de Jesús Rostro-Alanis, M., Muñoz-Gutiérrez, B. D., Iqbal, H. M., Kannan, S., & Parra-Saldívar, R. (2018). Emergent contaminants: Endocrine disruptors and their laccase-assisted degradation–a review. *Science of the Total Environment, 612*, 1516–1531.

Bhunia, S. K., & Jana, N. R. (2014). Reduced graphene oxide-silver nanoparticle composite as visible light photocatalyst for degradation of colorless endocrine disruptors. *ACS applied materials & interfaces, 6*(22), 20085–20092.

Borrirukwisitsak, S., Keenan, H. E., & Gauchotte-Lindsay, C. (2012). Effects of salinity, pH and temperature on the octanol-water partition coefficient of bisphenol A. *International Journal of Environmental Science and Development, 3*(5), 460.

Caliman, F. A., & Gavrilescu, M. (2009). Pharmaceuticals, personal care products and endocrine disrupting agents in the environment - A review. *Clean - Soil, Air, Water, 37*, 277–303. https://doi.org/10.1002/clen.200900038.

Choi, K. J., Kim, S. G., Kim, C. W., & Kim, S. H. (2005). Effects of activated carbon types and service life on removal of endocrine disrupting chemicals: Amitrol, nonylphenol, and bisphenol-A. *Chemosphere, 58*(11), 1535–1545.

Combarnous, Y., & Nguyen, T. M. D. (2019). Comparative overview of the mechanisms of action of hormones and endocrine disruptor compounds. *Toxics, 7*(1), 5.

Dai, R., Han, H., Wang, T., Li, X., & Wang, Z. (2021a). Enhanced removal of hydrophobic endocrine disrupting compounds from wastewater by nanofiltration membranes intercalated with hydrophilic MoS2 nanosheets: Role of surface properties and internal nanochannels. *Journal of Membrane Science, 628*, 119267. https://doi.org/10.1016/j.memsci.2021.119267.

Dai, R., Han, H., Zhu, Y., Wang, X., & Wang, Z. (2021b). Tuning the primary selective nanochannels of MOF thin-film nanocomposite nanofiltration membranes for efficient removal of hydrophobic endocrine disrupting compounds. *Frontiers in Environmental Science and Engineering, 16*, 40. https://doi.org/10.1007/s11783-021-1474-7.

Datel, J. V., & Hrabankova, A. (2020). Pharmaceuticals load in the Svihov water reservoir (Czech Republic) and impacts on quality of treated drinking water. *Water (Switzerland), 12.*https://doi.org/10.3390/W12051387.

Dharupaneedi, S. P., Nataraj, S. K., & Nadagouda, M. (2019). Membrane-based separation of potential emerging pollutants. *Separation and Purification Technology, 210*, 850–866. https://doi.org/10.1016/j.seppur.2018.09.003.

EFSA Scientific Committee. (2013). Scientific Opinion on the hazard assessment of endocrine disruptors: Scientific criteria for identification of endocrine disruptors and appropriateness of existing test methods for assessing effects mediated by these substances on human health and the environment. *EFSA Journal, 11*(3), 3132.

Eltoukhy, A., Jia, Y., Nahurira, R., Abo-Kadoum, M. A., Khokhar, I., Wang, J., & Yan, Y. (2020). Biodegradation of endocrine disruptor Bisphenol A by Pseudomonas putida strain YC-AE1 isolated from polluted soil, Guangdong, China. *BMC Microbiology, 20*, 1–14.

Felizeter, S., McLachlan, M.S., & De Voogt, P. (2014). Root uptake and translocation of perfluorinated alkyl acids by three hydroponically grown crops. *Journal of Agricultural and Food Chemistry, 62*, 3334–3342. https://doi.org/10.1021/jf500674j.

Fonseca, V. F., Lara, L. Z., Bertoldi, C. F., Waldman, W. R., Fernandes, A. N. (2024). Polyamide microplastics as endocrine disruptors: A study about the influence of photodegradation and sorption mechanisms under distinct environmental context. In *Microplastics and Pollutants: Interactions, Degradations and Mechanisms* (pp. 149–172). Cham: Springer Nature Switzerland.

Fu, J., Lee, W., Coleman, C., Nowack, K., Carter, J., & Huang, C. (2019). Removal of pharmaceuticals and personal care products by two-stage biofiltration for drinking water treatment. *Science of the Total Environment, 664*, 240–248. https://doi.org/10.1016/j.scitotenv.2019.02.026.

Gao, X., Kang, S., Xiong, R., & Chen, M. (2020). Environment-friendly removal methods for endocrine disrupting chemicals. *Sustainability, 12*(18), 7615.

Gasser, C. A., Ammann, E. M., Shahgaldian, P., & Corvini, P. F. X. (2014). Laccases to take on the challenge of emerging organic contaminants in wastewater. *Applied Microbiology and Biotechnology, 98*, 9931–9952.

Gobelius, L., Lewis, J., & Ahrens, L. (2017). Plant uptake of per-and polyfluoroalkyl substances at a contaminated fire training facility to evaluate the phytoremediation potential of various plant species. *Environmental Science & Technology, 51*(21), 12602–12610.

Gong, X., Huang, D., Liu, Y., Zeng, G., Wang, R., Wei, J., Huang, C., Xu, P., Wan, J., & Zhang, C. (2018). Pyrolysis and reutilization of plant residues after phytoremediation of heavy metals contaminated sediments: For heavy metals stabilization and dye adsorption. *Bioresource Technology, 253*, 64–71.

González-González, R. B., Parra-Arroyo, L., Parra-Saldívar, R., Ramirez-Mendoza, R. A., & Iqbal, H. M. N. (2022). Nanomaterial-based catalysts for the degradation of endocrine-disrupting chemicals—A way forward to environmental remediation. *Materials Letters, 308*, 131217. https://doi.org/10.1016/j.matlet.2021.131217

Gore, A. C., Chappell, V. A., Fenton, S. E., Flaws, J. A., Nadal, A., Prins, G. S., Toppari, J., & Zoeller, R. T. (2015). Executive summary to EDC-2: The endocrine society's second scientific statement on endocrine-disrupting chemicals. *Endocrine Reviews, 36*(6), 593–602.

Gunarathne, V., Gunatilake, S. R., Wanasinghe, S. T., Atugoda, T., Wijekoon, P., Biswas, J. K., Vithanage, M. (2020). Phytoremediation for E-waste contaminated sites. In *Handbook of electronic waste management* (pp. 141–170). Butterworth-Heinemann.

Guo, J., Dai, Y. Z., Chen, X. J., Zhou, L. L., & Liu, T. H. (2017). Synthesis and characterization of $Ag_3PO_4/LaCoO_3$ nanocomposite with superior mineralization potential for bisphenol A degradation under visible light. *Journal of Alloys and Compounds, 696*, 226–233.

Herath, I., & Vithanage, M. (2015). Phytoremediation in constructed wetlands. In *Phytoremediation: Management of Environmental Contaminants* (Vol. 2, pp. 243–263).

Huang, Y., Li, W., Qin, L., Xie, X., Gao, B., Sun, J., & Li, A. (2019). Distribution of endocrine-disrupting chemicals in colloidal and soluble phases in municipal secondary effluents and their removal by different advanced treatment processes. *Chemosphere, 219*, 730–739. https://doi.org/10.1016/j.chemosphere.2018.11.201.

Huang, D., Xiao, R., Du, L., Zhang, G., Yin, L., Deng, R., & Wang, G. (2021). Phytoremediation of poly-and perfluoroalkyl substances: A review on aquatic plants, influencing factors, and phytotoxicity. *Journal of Hazardous Materials, 418*, 126314.

Huerta-Fontela, M., Galceran, M. T., & Ventura, F. (2011). Occurrence and removal of pharmaceuticals and hormones through drinking water treatment. *Water Resources, 45*, 1432–1442. https://doi.org/10.1016/j.watres.2010.10.036.

Hwang, K. A., & Choi, K. C. (2015). Endocrine-disrupting chemicals with estrogenicity posing the risk of cancer progression in estrogen-responsive organs. In *Advances in Molecular Toxicology* (Vol. 9, pp. 1–33). Elsevier.

Ifelebuegu, A. O., & Ezenwa, C. P. (2011). Removal of endocrine disrupting chemicals in wastewater treatment by fenton-like oxidation. *Water, Air, & Soil Pollution, 217*, 213–220. https://doi.org/10.1007/s11270-010-0580-0.

Im, J. -K., Boateng, L. K., Flora, J. R. V., Her, N., Zoh, K. -D., Son, A., & Yoon, Y. (2014). Enhanced ultrasonic degradation of acetaminophen and naproxen in the presence of powdered activated carbon and biochar adsorbents. *Separation and Purification Technology, 123*, 96–105. https://doi.org/10.1016/j.seppur.2013.12.021.

Janošek, J., Hilscherová, K., Bláha, L., & Holoubek, I. (2006). Environmental xenobiotics and nuclear receptors—Interactions, effects and in vitro assessment. *Toxicology in Vitro, 20*(1), 18–37.

Jiang, J. Q., Zhou, Z., & Sharma, V. K. (2013). Occurrence, transportation, monitoring and treatment of emerging micro-pollutants in waste water—A review from global views. *Microchemical Journal, 110*, 292–300.

Jiang, Z., Feng, L., Zhu, J., Liu, B., Li, X., Chen, Y., & Khan, S. (2021). Construction of a hierarchical NiFe2O4/CuInSe2 (p-n) heterojunction: Highly efficient visible-light-driven photocatalyst in the degradation of endocrine disruptors in an aqueous medium. *Ceramics International, 47*, 8996–9007. https://doi.org/10.1016/j.ceramint.2020.12.022.

Jones, S. M., & Solomon, E. I. (2015). Electron transfer and reaction mechanism of laccases. *Cellular and Molecular Life Sciences, 72*, 869–883.

Kabir, E. R., Rahman, M. S., & Rahman, I. (2015). A review on endocrine disruptors and their possible impacts on human health. *Environmental Toxicology and Pharmacology, 40*(1), 241–258.

Kamaraj, M., Ranjith, K. S., Sivaraj, R., RT, R. K., & Salam, H. A. (2014). Photocatalytic degradation of endocrine disruptor Bisphenol-A in the presence of prepared $Ce_xZn_{1-x}O$ nanocomposites under irradiation of sunlight. *Journal of Environmental Sciences, 26*(11), 2362–2368.

Kidd, K. A., Schindler, D. W., Muir, D. C., Lockhart, W. L., & Hesslein, R. H. (1995). High concentrations of toxaphene in fishes from a subarctic lake. *Science, 269*(5221), 240–242.

Kim, Y. J., & Nicell, J. A. (2006). Impact of reaction conditions on the laccase-catalyzed conversion of bisphenol A. *Bioresource Technology, 97*, 1431–1442. https://doi.org/10.1016/j.biortech.2005.06.017.

Kinney, C. A., Furlong, E. T., Werner, S. L., & Cahill, J. D. (2006). Presence and distribution of wastewater-derived pharmaceuticals in soil irrigated with reclaimed water. *Environmental Toxicology and Chemistry, 25*, 317–326. https://doi.org/10.1897/05-187R.1.

Koyuncu, I., Arikan, O. A., Wiesner, M. R., & Rice, C. (2008). Removal of hormones and antibiotics by nanofiltration membranes. *Journal of Membrane Science, 309*, 94–101. https://doi.org/10.1016/j.memsci.2007.10.010.

Kurwadkar, S., Hoang, T. V., Malwade, K., Kanel, S. R., Harper, W. F., & Struckhoff, G. (2019). Application of carbon nanotubes for removal of emerging contaminants of concern in engineered water and wastewater treatment systems. *Nanotechnology and Environmental Engineering, 1*. https://doi.org/10.1007/s41204-019-0059-1.

Kwon, J. W., & Armbrust, K. L. (2008). Aqueous solubility, n-octanol-water partition coefficient, and sorption of five selective serotonin reuptake inhibitors to sediments and soils. *Bulletin of Environmental Contamination and Toxicology, 81*, 128–135. https://doi.org/10.1007/s00128-008-9401-1.

La Rocca, C., & Mantovani, A. (2006). From environment to food: The case of PCB. *Annali-Istituto Superiore di Sanita, 42*(4), 410.

Lam, S. M., Sin, J. C., Abdullah, A. Z., & Mohamed, A. R. (2013). Efficient photodegradation of endocrine-disrupting chemicals with Bi_2O_3–ZnO nanorods under a compact fluorescent lamp. *Water, Air, & Soil Pollution, 224*, 1–11.

Laurenson, J. P., Bloom, R. A., Page, S., & Sadrieh, N. (2014). Ethinyl estradiol and other human pharmaceutical estrogens in the aquatic environment: A review of recent risk assessment data. *The AAPS Journal, 16*, 299–310.

Lauretta, R., Sansone, A., Sansone, M., Romanelli, F., & Appetecchia, M. (2019). Endocrine disrupting chemicals: Effects on endocrine glands. *Frontiers in Endocrinology, 10*, 178.

Liu, J., Zhou, J., Wu, Z., Tian, X., An, X., Zhang, Y., Zhang, G., Deng, F., Meng, X., & Qu, J. (2022). Concurrent elimination and stepwise recovery of Pb (II) and bisphenol A from water using β–cyclodextrin modified magnetic cellulose: Adsorption performance and mechanism investigation. *Journal of Hazardous Materials, 432*, 128758. https://doi.org/10.1016/j.jhazmat.2022.128758.

Loffredo, E., Picca, G., & Parlavecchia, M. (2021). Single and combined use of Cannabis sativa L. and carbon-rich materials for the removal of pesticides and endocrine-disrupting chemicals from water and soil. *Environmental Science and Pollution Research, 28*, 3601–3616.

Macedo, S., Teixeira, E., Gaspar, T. B., Boaventura, P., Soares, M. A., Miranda-Alves, L., & Soares, P. (2023). Endocrine-disrupting chemicals and endocrine neoplasia: A forty-year systematic review. *Environmental Research, 218*, 114869.

Mackay, D., & Fraser, A. (2000). Bioaccumulation of persistent organic chemicals: Mechanisms and models. *Environmental Pollution, 110*(3), 375–391.

Madhavi, V., & Lele, S. S. (2009). Laccase: properties and applications. *BioResources, 4*(4).

Mallakpour, S., & Soltanian, S. (2016). Surface functionalization of carbon nanotubes: fabrication and applications. *RSC Advances, 6*, 109916–109935. https://doi.org/10.1039/C6RA24522F.

Mantovani, A. (2016). Endocrine disrupters and the safety of food chains. *Hormone Research in Paediatrics, 86*(4), 279–288.

Mantovani, A., Frazzoli, C., & La Rocca, C. (2009). Risk assessment of endocrine-active compounds in feeds. *The Veterinary Journal, 182*, 392–401. https://doi.org/10.1016/j.tvjl.2008.08.005.

Mantovani, A., Maranghi, F., La Rocca, C., Tiboni, G.M., & Clementi, M. (2008). The role of toxicology to characterize biomarkers for agrochemicals with potential endocrine activities. *Reproductive Toxicology, 26*, 1–7. https://doi.org/10.1016/j.reprotox.2008.05.063.

Martín-Lara, M. A., Calero, M., Ronda, A., Iáñez-Rodríguez, I., & Escudero, C. (2020). Adsorptive behavior of an activated carbon for bisphenol A removal in single and binary (bisphenol A— heavy metal) solutions. *Water, 12*, 2150. https://doi.org/10.3390/w12082150.

Mayakaduwage, S., Ekanayake, A., Kurwadkar, S., Rajapaksha, A. U., & Vithanage, M. (2022). Phytoremediation prospects of per-and polyfluoroalkyl substances: A review. *Environmental Research, 212*, 113311.

Mnif, W., Hassine, A. I. H., Bouaziz, A., Bartegi, A., Thomas, O., & Roig, B. (2011). Effect of endocrine disruptor pesticides: A review. *International Journal of Environmental Research and Public Health, 8*(6), 2265–2303.

Monneret, C. (2017). What is an endocrine disruptor? *Comptes Rendus. Biologies, 340*(9–10), 403–405.

Mudumbi, J. B. N., Ntwampe, S. K. O., Muganza, M., & Okonkwo, J. O. (2014). Susceptibility of riparian wetland plants to perfluorooctanoic acid (PFOA) accumulation. *International Journal of Phytoremediation, 16*(9), 926–936.

Naddeo, V., Landi, M., Scannapieco, D., & Belgiorno, V. (2013). Sonochemical degradation of twenty-three emerging contaminants in urban wastewater. *Desalination and Water Treatment, 51*, 37–41. https://doi.org/10.1080/19443994.2013.769696.

Noszczyńska, M., Chodór, M., Jałowiecki, Ł, & Piotrowska-Seget, Z. (2021). A comprehensive study on bisphenol A degradation by newly isolated strains Acinetobacter sp. K1MN and Pseudomonas sp. BG12. *Biodegradation, 32*, 1–15.

Noutsopoulos, C., Mamais, D., Mpouras, T., Kokkinidou, D., Samaras, V., Antoniou, K., & Gioldasi, M. (2014). The role of activated carbon and disinfection on the removal of endocrine disrupting chemicals and non-steroidal anti-inflammatory drugs from wastewater. *Environmental Technology (United Kingdom), 35*, 698–708. https://doi.org/10.1080/09593330.2013.846923.

Ojajuni, O., Saroj, D., & Cavalli, G. (2015). Removal of organic micropollutants using membrane-assisted processes: A review of recent progress. *Environmental Technology Reviews, 4*, 37–41. https://doi.org/10.1080/21622515.2015.1036788.

Panja, S., Sarkar, D., & Datta, R. (2020). Removal of tetracycline and ciprofloxacin from wastewater by vetiver grass (Chrysopogon zizanioides (L.) Roberty) as a function of nutrient concentrations. *Environmental Science and Pollution Research, 27*(28), 34951–34965.

Pant, H. R., Park, C. H., Pant, B., Tijing, L. D., Kim, H. Y., & Kim, C. S. (2012). Synthesis, characterization, and photocatalytic properties of ZnO nano-flower containing TiO_2 NPs. *Ceramics International, 38*(4), 2943–2950.

Park, J. S., Her, N., & Yoon, Y. (2011). Ultrasonic degradation of bisphenol A, 17β-estradiol, and 17α-ethinyl. *Desalination and Water Treatment, 30*(1–3), 300–309.

Patel, M., Kumar, R., Kishor, K., Mlsna, T., Pittman, C. U., & Mohan, D. (2019). Pharmaceuticals of emerging concern in aquatic systems: Chemistry, occurrence, effects, and removal methods. *Chemical Reviews, 119*, 3510–3673. https://doi.org/10.1021/acs.chemrev.8b00299.

Pi, N., Ng, J. Z., & Kelly, B. C. (2017). Bioaccumulation of pharmaceutically active compounds and endocrine disrupting chemicals in aquatic macrophytes: Results of hydroponic experiments with Echinodorus horemanii and Eichhornia crassipes. *Science of the Total Environment, 601–602*, 812–820. https://doi.org/10.1016/j.scitotenv.2017.05.137.

Pirro, V., Girolami, F., Spalenza, V., Gardini, G., Badino, P., & Nebbia, C. (2015). Set-up of a multivariate approach based on serum biomarkers as an alternative strategy for the screening evaluation of the potential abuse of growth promoters in veal calves. *Food Additives & Contaminants: Part A, 32*(5), 702–711.

Pocar, P., Grieco, V., Aidos, L., & Borromeo, V. (2023). Endocrine-disrupting chemicals and their effects in pet dogs and cats: An overview. *Animals, 13*(3), 378.

Rao, A., Kumar, A., Dhodapkar, R., & Pal, S. (2021). Adsorption of five emerging contaminants on activated carbon from aqueous medium: Kinetic characteristics and computational modeling for plausible mechanism. *Environmental Science and Pollution Research, 28*, 21347–21358.

Rocha, P. R. S., Oliveira, V. D., Vasques, C. I., Dos Reis, P. E. D., & Amato, A. A. (2021). Exposure to endocrine disruptors and risk of breast cancer: A systematic review. *Critical Reviews in Oncology/hematology, 161*, 103330.

Rodgers, K. M., Udesky, J. O., Rudel, R. A., & Brody, J. G. (2018). Environmental chemicals and breast cancer: An updated review of epidemiological literature informed by biological mechanisms. *Environmental Research, 160*, 152–182.

Rosenblum, E. R., Stauber, R. E., Van Thiel, D. H., Campbell, I. M., Gavaler, J. S. (1993). Assessment of the estrogenic activity of phytoestrogens isolated from bourbon and beer. *Alcoholism: Clinical and Experimental Research, 17*(6), 1207–1209.

Ruhí i Vidal, A., Acuña i Salazar, V., Barceló i Cullerés, D., Huerta Buitrago, B., Mor Roy, J. R., Rodríguez Mozaz, S., & Sabater, S. (2016). Bioaccumulation and trophic magnification of pharmaceuticals and endocrine disruptors in a Mediterranean river food web. ©*Science of the Total Environment, 540*, 250–259.

Schug, T. T., Janesick, A., Blumberg, B., & Heindel, J. J. (2011). Endocrine disrupting chemicals and disease susceptibility. *The Journal of Steroid Biochemistry and Molecular Biology, 127*(3–5), 204–215.

Silva, L. G. R., Costa, E. P., Starling, M. C. V. M., dos Santos Azevedo, T., Bottrel, S. E. C., Pereira, R. O., Sanson, A. L., Afonso, R. J. C. F., & Amorim, C. C. (2021). LED irradiated photo-Fenton for the removal of estrogenic activity and endocrine disruptors from wastewater treatment plant effluent. *Environmental Science and Pollution Research, 28*, 24067–24078.

Slama, R., Bourguignon, J. P., Demeneix, B., Ivell, R., Panzica, G., Kortenkamp, A., & Zoeller, R. T. (2016). Scientific issues relevant to setting regulatory criteria to identify endocrine-disrupting substances in the European Union. *Environmental Health Perspectives, 124*(10), 1497–1503.

Swedenborg, E., Rüegg, J., Mäkelä, S., & Pongratz, I. (2009). Endocrine disruptive chemicals: Mechanisms of action and involvement in metabolic disorders. *Journal of Molecular Endocrinology, 43*(1), 1–10.

Tabb, M. M., & Blumberg, B. (2006). New modes of action for endocrine-disrupting chemicals. *Molecular Endocrinology, 20*(3), 475–482.

Tanaka, T., Tonosaki, T., Nose, M., Tomidokoro, N., Kadomura, N., Fujii, T., & Taniguchi, M. (2001). Treatment of model soils contaminated with phenolic endocrine-disrupting chemicals with laccase from Trametes sp. in a rotating reactor. *Journal of bioscience and bioengineering, 92*(4), 312–316.

Thathola, P., Agnihotri, V., Pandey, A., & Upadhyay, S. K. (2022). Biodegradation of bisphenol A using psychrotolerant bacterial strain Pseudomonas palleroniana GBPI_508. *Archives of Microbiology, 204*(5), 272.

Toppari, J. (2008). Environmental endocrine disrupters. *Sexual Development, 2*(4–5), 260–267.

Tursi, A., Chatzisymeon, E., Chidichimo, F., Beneduci, A., & Chidichimo, G. (2018). Removal of endocrine disrupting chemicals from water: Adsorption of bisphenol-a by biobased hydrophobic functionalized cellulose. *International Journal of Environmental Research and Public Health, 15.* https://doi.org/10.3390/ijerph15112419.

Vela, N., Calín, M., Yáñez-Gascón, M. J., Garrido, I., Pérez-Lucas, G., Fenoll, J., & Navarro, S. (2018). Photocatalytic oxidation of six endocrine disruptor chemicals in wastewater using ZnO at pilot plant scale under natural sunlight. *Environmental Science and Pollution Research, 25,* 34995–35007. https://doi.org/10.1007/s11356-018-1716-9.

Villemur, R., Dos Santos, S. C. C., Ouellette, J., Juteau, P., Lépine, F., & Déziel, E. (2013). Biodegradation of endocrine disruptors in solid-liquid two-phase partitioning systems by enrichment cultures. *Applied and Environmental Microbiology, 79*(15), 4701–4711.

Wang, Y., Wang, X., Li, M., Dong, J., Sun, C., & Chen, G. (2018). Removal of pharmaceutical and personal care products (PPCPs) from municipal wastewater with integrated membrane systems, MBR-RO/NF. *International Journal of Environmental Research and Public Health, 15.* https://doi.org/10.3390/ijerph15020269.

Wu, Q., Shi, H., Adams, C. D., Timmons, T., & Ma, Y. (2012). Oxidative removal of selected endocrine-disruptors and pharmaceuticals in drinking water treatment systems, and identification of degradation products of triclosan. *Science of the Total Environment, 439,* 18–25. https://doi.org/10.1016/j.scitotenv.2012.08.090.

Xie, X., Lü, W., & Chen, X. (2013). Binding of the endocrine disruptors 4-tert-octylphenol and 4-nonylphenol to human serum albumin. *Journal of Hazardous Materials, 248,* 347–354.

Xie, Y., Guo, X., Liang, Z., & Shim, H. (2023). Biochemical pathways and enhanced degradation of endocrine disruptor di-2-ethylhexyl phthalate by an indigenous isolate Bacillus sp. MY156. *International Biodeterioration & Biodegradation, 176,* 105523. https://doi.org/10.1016/j.ibiod.2022.105523.

Xu, B., Gao, N., Rui, M., Wang, H., & Wu, H. (2006). Degradation of endocrine disruptor bisphenol A in drinking water by ozone oxidation. *Huan jing ke xue = Huanjing Kexue, 27,* 294–299.

Yoon, K., Kwack, S. J., Kim, H. S., & Lee, B. M. (2014). Estrogenic endocrine-disrupting chemicals: Molecular mechanisms of actions on putative human diseases. *Journal of Toxicology and Environmental Health, Part B, 17*(3), 127–174.

Zamri, M. F. M. A., Bahru, R., Suja, F., Shamsuddin, A. H., Pramanik, S. K., & Fattah, I. M. R. (2021). Treatment strategies for enhancing the removal of endocrine-disrupting chemicals in water and wastewater systems. *Journal of Water Process Engineering, 41.* https://doi.org/10.1016/j.jwpe.2021.102017.

Zatloukalová, K., Obalová, L., Koci, K., Capek, L., Matej, Z., Snajdhaufova, H., Ryczkowski, J., & Slowik, G. (2017). Photocatalytic degradation of endocrine disruptor compounds in water over immobilized TiO_2 photocatalysts.

Zhang, Y. Z., Meng, W., & Zhang, Y. (2014). Occurrence and partitioning of phenolic endocrine-disrupting chemicals (EDCs) between surface water and suspended particulate matter in the North Tai Lake Basin, Eastern China. *Bulletin of Environmental Contamination and Toxicology, 92,* 148–153. https://doi.org/10.1007/s00128-013-1136-y.

Zhang, Y., Zeng, Z., Zeng, G., Liu, X., Liu, Z., Chen, M., Liu, L., Li, J., & Xie, G. (2012). Effect of Triton X-100 on the removal of aqueous phenol by laccase analyzed with a combined approach of experiments and molecular docking. *Colloids and Surfaces b: Biointerfaces, 97,* 7–12.

Zhang, Q., Xue, C., Owens, G., & Chen, Z. (2022a). Isolation and identification of 17β-estradiol degrading bacteria and its degradation pathway. *Journal of Hazardous Materials, 423*, 127185.

Zhang, C., Wu, J., Chen, Q., Tan, H., Huang, F., Guo, J., Zhang, X., Yu, H., & Shi, W. (2022b). Allosteric binding on nuclear receptors: Insights on screening of non-competitive endocrine-disrupting chemicals. *Environment International, 159*, 107009.

Zhao, C., Zhang, G., & Jiang, J. (2021). Enhanced phytoremediation of bisphenol a in polluted lake water by seedlings of ceratophyllum demersum and myriophyllum spicatum from in vitro culture. *International Journal of Environmental Research and Public Health, 18*(2), 810.